ANIMAL AND PLANT
Anatomy

11

Kingdoms of life – Indexes

 Marshall Cavendish
Reference
New York

CONTRIBUTORS

Roger Avery; Richard Beatty; Amy-Jane Beer; Erica Bower; Trevor Day; Erin Dolan; Bridget Giles; Natalie Goldstein; Tim Harris; Christer Hogstrand; Rob Houston; John Jackson; Tom Jackson; James Martin; Chris Mattison; Katie Parsons; Ray Perrins; Kieran Pitts; Adrian Seymour; Steven Swaby; John Woodward.

CONSULTANTS

Barbara Abraham, Hampton University, VA; Glen Alm, University of Guelph, Ontario, Canada; Roger Avery, Bristol University, England; Amy-Jane Beer, University of London, England; Deborah Bodolus, East Stroudsburg University, PA; Allan Bornstein, Southeast Missouri State University, MO; Erica Bower, University of London, England; John Cline, University of Guelph, Ontario, Canada; Trevor Day, University of Bath, England; John Friel, Cornell University, NY; Valerius Geist, University of Calgary, Alberta, Canada; John Gittleman, University of Virginia, VA; Tom Jenner, Academia Británica Cuscatleca, El Salvador; Bill Kleindl, University of Washington, Seattle, WA; Thomas Kunz, Boston University, MA; Alan Leonard, Florida Institute of Technology, FL; Sally-Anne Mahoney, Bristol University, England; Chris Mattison; Andrew Methven, Eastern Illinois University, IL; Graham Mitchell, King's College, London, England; Richard Mooi, California Academy of Sciences, San Francisco, CA; Ray Perrins, Bristol University, England; Kieran Pitts, Bristol University, England; Adrian Seymour, Bristol University, England; David Spooner, University of Wisconsin, WI; John Stewart, Natural History Museum, London, England; Erik Terdal, Northeastern State University, Broken Arrow, OK; Phil Whitfield, King's College, University of London, England.

Marshall Cavendish

99 White Plains Road
Tarrytown, NY 10591–9001

www.marshallcavendish.us

© 2007 Marshall Cavendish Corporation

Library of Congress Cataloging-in-Publication Data
Animal and plant anatomy.
      p. cm.
   ISBN-13: 978-0-7614-7662-7 (set: alk. paper)
   ISBN-10: 0-7614-7662-8 (set: alk. paper)
   ISBN-13: 978-0-7614-7675-7 (vol. 11)
   ISBN-10: 0-7614-7675-X (vol. 11)
   1. Anatomy. 2. Plant anatomy. I. Marshall Cavendish Corporation. II.
Title.

   QL805.A55 2006
   571.3--dc22

                                                   2005053193

Printed in China
09 08 07 06 1 2 3 4 5

MARSHALL CAVENDISH

Editor: Joyce Tavolacci
Editorial Director: Paul Bernabeo
Production Manager: Mike Esposito

THE BROWN REFERENCE GROUP PLC

Project Editor: Tim Harris
Deputy Editor: Paul Thompson
Subeditors: Jolyon Goddard, Amy-Jane Beer, Susan Watts
Designers: Bob Burroughs, Stefan Morris
Picture Researchers: Susy Forbes, Laila Torsun
Indexer: Kay Ollerenshaw
Illustrators: The Art Agency, Mick Loates, Michael Woods
Managing Editor: Bridget Giles

Contents

Volume 1

Albatross 10
Amoeba 20
Ant 26
Apple tree 36
Bacteria 46
Bullfrog 54
Cell biology and
  genetics 64
Chimpanzee 88
Circulatory system 106
Coelacanth 122
Crab 132

Volume 2

Crocodile 148
Diatom 164
Digestive and
  excretory systems 172
Dolphin 188
Dragonfly 208
Eagle 220
Earthworm 234
Elephant 244
Endocrine and
  exocrine systems 264
Fern 276

Volume 3

Fruit bat 292
Giant anteater 312
Giant centipede 332

Giant clam 342
Giraffe 356
Gray whale 374
Green anaconda 396
Grizzly bear 410

Volume 4

Gulper eel 436
Hagfish 446
Hammerhead shark 460
Hare 476
Hawkmoth 494
Hippopotamus 506
Honeybee 526
Housefly 538
Human 550

Volume 5

Hummingbird 580
Hyena 592
Immune, defense, and
  lymphatic systems 608
Jackson's chameleon 624
Jellyfish 636
Kangaroo 648
Lion 664
Lobster 684
Louse 692
Manatee 700

Volume 6

Mandrill 724
Marsh grass 744
Muscular system 754
Mushroom 770
Nervous system 780
Newt 800
Octopus 812
Orchid 824
Ostrich 834
Otter 844

Volume 7

Penguin 868
Platypus 880
Porcupine 898
Potato 916
Puma 928
Rat 946
Red deer 964
Reproductive system 984

Volume 8

Respiratory system 1012
Rhinoceros 1032
Saguaro cactus 1048
Sailfish 1060
Scorpion 1074
Sea anemone 1084
Sea horse 1096
Seal 1110

Seaweed 1126
Sequoia 1138

Volume 9

Skeletal system 1156
Sloth 1176
Snapping turtle 1194
Squirrel 1208
Starfish 1224
Stingray 1238
Tarantula 1252
Tortoise 1266
Trout 1280

Volume 10

Venus flytrap 1300
Virus 1314
Weasel 1324
Weevil 1344
Wildebeest 1360
Wolf 1382
Woodpecker 1404
Zebra 1418

Volume 11

Kingdoms of life 1444
The study
  of anatomy 1456
Glossary 1462
Resources for
  further study 1472
Indexes 1480

Kingdoms of life

Anatomy is a branch of biology, and biology is the science of living things. People who study biology do so for many reasons, but high on almost everyone's list is a deep fascination with the complexity and variety of life. It seems that the more people discover about living organisms, the more we realize there is still to know.

The classification, or taxonomy, of life-forms is the science of organizing living things into categories based on their relationships to each other. Taxonomy is an indispensable part of biology as a whole.

The diversity of life

The organisms that appear within the pages of *Animal and Plant Anatomy* come in a startling array of shapes and sizes, from protists too small to see except with a powerful microscope to some of the largest animals alive, including elephants and whales. The volumes in the set look closely at the bodies of bullfrogs and dragonflies, clams and chimpanzees, and compare the internal workings of bacteria, fungi, plants, and other life-forms with those of our own bodies. Even so, this encyclopedia barely scratches the surface. To consider life in all its forms is not only too big an undertaking for one book, but too big for all the books biologists have ever written. This is because there are species of animals, plants, fungi, algae, protists, and bacteria that no one has ever yet seen, let alone described scientifically.

▼ *The beetle* Calsoma sycophanta *feeding on the caterpillar of a gypsy moth. Entomologists (scientists who study insects) know of more than 350,000 species of beetles, but the true total may be several times larger than this number.*

How many species are there?

Even to put a number to the living species that have already been named and described is a bit of a guessing game. The figure is probably in the region of 1.7 million, but there is no single list of species that provides a definitive total. Such lists are sometimes attempted for small groups of organisms—for example, birds, crustaceans, or orchids. However, even these nearly always disagree with similar lists that have gone before or come after, because biologists are always discovering new species or tinkering with the classification of others. Biologists often divide life-forms into new groups or lump them together when they realize they are essentially the same. And, sadly, species are becoming extinct at an alarming rate. However, despite difficulties in coming up with a good estimate of the number of species, what everyone does agree is that the number of known species is far outweighed by the number of those yet to be discovered. Estimates for the total number of visible species (that is, excluding bacteria and protists) vary from a conservative 8 million to more

Estimating diversity

In an attempt to gain an understanding of the diversity of species, the American biologist Terry Erwin performed an experiment in the tropical forests of Peru in the early 1980s. Using a technique called fogging, he treated a small clump of trees (all of one species) with a fine mist of powerful but biodegradable insecticide. Invertebrates living in the trees died and fell from the canopy, and Erwin collected them. The majority were species new to science, and they included 1,100 species of beetles alone. From this experiment, Erwin used a series of simple but controversial calculations to come up with an estimate for total global biodiversity of 30 million species. Many people thought this was an absurd overestimate. In fact, it may turn out to be conservative.

than 100 million. Bringing order to these life-forms is the task of the taxonomist.

What is a species?

Before tackling the question of how many species there are, it is important to have a firm grasp of what makes a species. The traditionally accepted definition of biological

◀ *Early botanists studied orchids, such as this* Cattleya *species, for their distinctive and beautiful flowers. In the mid-18th century the Swedish biologist Linnaeus recognized eight orchid genera, or groups of closely related species. Contemporary estimates put the total number of orchid species at about 20,000, in 850 genera.*

species was set down by the accomplished German-American biologist Ernst Mayr (1904–2005) in 1949. Mayr stated that a species is a group of organisms that can reproduce sexually to produce viable offspring. By "viable," Mayr meant offspring that were at least as fit for life as their parents and able to reproduce successfully themselves. Organisms that come from different species do not usually attempt to breed, but if crossbreeding should occur, and the hybrid survives, it will usually be congenitally weaker than either of its parents or, as in the familiar example of a female horse and male ass breeding to produce a mule, it will be infertile.

The barrier to successful breeding between species can take many forms. In many cases mating is physically impossible, and in most animal examples there is also a behavioral barrier, so that even if individuals from different species look similar they do not behave in a way the other species recognizes. For example, similar-looking birds may have very different songs and sexual displays, so individuals of one species will not be attracted to those of the other. However, if these barriers are breached and mating does occur, there are other obstacles to success. For example, the zygote (fertilized egg) may fail to divide properly; or the developing embryo may suffer a fatal developmental defect.

Despite all these hurdles, hybrids do sometimes occur. Usually, as with the mule, they are infertile. Mules cannot breed with other mules or any other animal, so it is not possible for a mule population to persist more than one generation. Hybrid animals are not always infertile, however. For example, on the Swedish island of Gotland two species of birds, called collared flycatchers and pied flycatchers, often hybridize. Sometimes the male offspring of mixed matings are fertile, although the females are probably always infertile. Hybrid animals are often less well adapted for life in the environment than their parents, so even if they do manage to breed, they will be less successful than either parent species, and thus they are ultimately doomed.

Where the ranges of two similar species meet there is sometimes a hybrid zone, in which a population of mixed-blood individuals is produced and continually topped up by accidental mating between the parent species. However, the hybrids never spread very far into the range of either parent species.

Sometimes the barrier is simply a matter of geography: two species that live in different places cannot meet and breed. Geographic separation is one of the prime factors in the evolution of new species. Given time, if a population of animals of one species is split in two (for example, on either side of a widening ocean or mountain range), characteristics will develop and be shaped differently by natural selection until the populations can be regarded as separate species.

Speciation and radiation

When, through the sheer good luck of a beneficial mutation, a group of animals hit on a new and successful way of doing things, or when they suddenly find themselves in an environment where competition is reduced (such as being blown off course onto a remote island not populated by any similar animals), they usually go though a rapid phase of diversification. The biologist and explorer Charles Darwin (1809–1882) observed the results of such a diversification (called a radiation) when he visited the Galápagos Islands. At some point in the past a small founding population of finches arrived on one

▼ Marine iguanas live only on the Galápagos Islands. The evolution of these reptiles is a mystery; although they are believed to have evolved about 8 million years ago, the oldest island in the Galápagos archipelago is only 4 million years old.

▲ *The fossilized remains of a mammoth, an extinct relative of elephants. By studying the anatomy of long-extinct fossil species, paleontologists can determine their relationship with animals that live today.*

of the islands. These first individuals and their offspring thrived. In time they spread to other islands where, in isolation from their relatives, they adapted to slightly different conditions and developed different specializations. By the time Darwin visited, radiations had progressed to a point where he was able to count 13 species of different, but related, birds on different islands.

The Cambrian "explosion"

The Cambrian explosion is the name given to a particularly large radiation that took place about 590 million years ago. It was a period of great diversification, when many of the major animal groups, or phyla, known today first put in an appearance. The arthropods, mollusks, and echinoderms all trace their origins to roughly to same period, but they were by no means alone.

The Cambrian explosion also gave rise to a great number of other groups—animals with bodies unlike anything that survives today. Speciation happened at an unprecedented rate, but as environmental niches were filled and competition became more intense, natural selection began ruthlessly eliminating all but the best-adapted new life-forms. The same process has happened over and over again throughout the history of life on Earth.

The Cambrian explosion took place at a time when all animals lived in the sea and few had hard body parts. Such animals do not fossilize easily, and it is only thanks to a generous twist of fate that we know of their existence. Around 530 million years ago a mud slide buried a whole coastal ecosystem: organisms of all shapes and sizes were buried alive in sediment so fine that it preserved tiny details of their delicate bodies. Over time the sediment turned to rock called shale, and what was once seafloor become dry land, then high mountains—the Canadian Rockies. The mountains were eroded, and the fine, layered shale was exposed. When, in 1909, geologists began investigating and examining these layers, they found a treasure trove of fossils

IN FOCUS

Animals of the Burgess Shale

One of the fossils found in the Burgess Shale in the Canadian Rockies was a wormlike creature called *Pikaia*, which is now believed to be one of the oldest known chordates. Some of the fossils are of animals with no known relatives, including *Haplophrentis;* some scientists believe it to be a mollusk, but others disagree. Most of the fossils are arthropods, including many trilobites, but there are also microfossils, including bacteria, cyanobacteria, dinoflagellates, and protists.

containing the preserved likenesses of some of the strangest animals imaginable. The discovery of the Burgess Shale fossils is credited to Charles Walcott (1850–1927), Director of the U. S. Geological Survey, but interpretation of the weird and wonderful species within the shale has become the lifework of many individuals, including Harry Whittington, Simon Conway Morris, and Derek Briggs.

The history of taxonomy

People have been classifying animals for millennia. Early classifications were mostly functional. For example, people could have drawn distinctions between animals that live in water and those that live on land, or between those that were good to eat and those that were dangerous, those that ran and crawled on land, and those that flew. The Greek philosopher Aristotle (384–322 BCE) was the first person to attempt a formal classification, and he separated living things into two kingdoms: animals and plants. He then classified the animals he knew as red-blooded or lacking blood, effectively separating what we now think of as vertebrates (mammals, birds, reptiles, amphibians, and fish) from invertebrates (including arthropods, worms, mollusks, and echinoderms).

The true father of taxonomy in its modern sense was the Swedish biologist Carl von Linné (1707–1778), also called Carolus Linnaeus. Linnaeus came up with a system for naming and grouping species in such a way that names and affiliations were much easier to grasp. Linnaeus gave each species two names: the first indicated what sort of organism it was, and the second identified its specific type. So the wolf became *Canis lupus*, and the domestic dog was *Canis familiaris*. From these names alone it would be possible for someone who had never seen a wolf to understand that is was in some way similar to a dog. And the Linnaean classification also solved the problem of coming up with enough names, since the specific part of the names could be used many times for different animals and plants without causing confusion. For example, the European red squirrel (*Sciurus vulgaris*), the herb self-heal (*Prunella vulgaris*), and the common wasp (*Vespula vulgaris*) are three very different organisms with the specific name *vulgaris*, but

there is no danger of confusing them because of their different generic names: *Sciurus*, *Prunella*, and *Vespula*. Incidentally, *vulgaris* means "common."

Linnaeus also came up with a hierarchy of groups into which the species he named would fit. Species were grouped into genera, genera into orders, and orders into classes. He followed Aristotle in grouping his classes into two kingdoms, plants (kingdom Plantae) and animals (kingdom Animalia).

Two additional levels were added later for convenience: families were fitted in between orders and genera, and the French biologist Georges Cuvier (1769–1832) established a category between class and kingdom. Cuvier called these major divisions "embranchements," but they soon became known as phyla, from the Greek for "tribe." Modern biologists also use a large number of intermediate levels such as subfamilies and superorders to organize the content of groups, but they are not always agreed on by everyone.

By the mid-19th century, however, naturalists realized that the two-kingdom system did not work. The single-cell organisms

▼ *A Eurasian red squirrel feeding on the seeds of a pine cone. This squirrel, which lives in mature coniferous and deciduous forests in Europe and Asia, belongs in the genus* Sciurus *with many other squirrels. Common membership in a genus denotes strong similarities in the animals' anatomy.*

such as those discovered in 1676 by the Dutch microscopist Antoni van Leeuwenhoek (1632–1723) (and now known to be bacteria) did not fit easily into either the plants or the animals. In 1866 the German biologist Ernst Haeckel created a third kingdom, which he called Protista (meaning "first created beings"), to accommodate the single-cell organisms. However, this system still left the fungi sitting rather uncomfortably alongside the plants.

The tree of life

Up until the late 19th century, systems for classifying the forms of life were basically hierarchical, with simple beings at the bottom of a ladder and humans at the top, at the very pinnacle of Creation (as it was then perceived to be). However, this view was dealt a devastating blow in 1859, with the publication by Charles Darwin of *On the Origin of Species*. Darwin's theory of evolution by natural selection held that all living species represent pinnacles in their own right: each one sits at the tip of a branch on a single, ancient, and colossal tree of life, like a family tree on the grandest scale.

Darwin was not the first person to come up with a theory of evolution. Others before him had noticed the way that some organisms appear to be closely related to others, and that they appeared to fall into natural categories. It

was generally accepted that in classifying organisms, biologists were not just making them easier to record and keep tabs on, but were following some kind of natural order in which furry animals that nourished their young on milk from glands in the skin fell into one group (mammals), and animals with six legs and a jointed exoskeleton fell into another (insects). It was clear to the naturalists of the day that a bird, with its warm blood and internal skeleton, was more similar to a mammal, and that a shrimp allied itself more closely with the insects. However, it was Darwin who unlocked the deeper significance of these relationships, showing that the differences between species could have evolved by a process of natural selection and that similarities between related organisms are therefore an indication of their shared ancestry.

By the mid-20th century the discovery of bacteria had led biologists to recognize that there were some divisions of life more fundamental than the differences among protists, plants, and animals. A close look at the fine structure of bacterial cells revealed that they were organized differently from those of other life-forms. Instead of containing organelles, such as the nucleus and mitochondria, packaged neatly in membranes, bacterial cells were much simpler—with contents that mingled freely. Organisms with

this simple cell structure became known as prokaryotes, while those with membrane-bound organelles, including plants, animals, fungi, and protists, were called eukaryotes. The tree of life was then redrawn by the American Robert Whittaker (1920–1980) to show five kingdoms: Animalia, Plantae, Fungi, Protista, and Monera (bacteria). The system gained wide acceptance in the scientific community during the 1980s thanks to the popular book *Five Kingdoms* by Lynn Margulis.

Cladistics

At around the same time that the five-kingdom system of classification was being proposed and accepted, a German scientist named Willi Hennig (1913–1976) was formalizing the idea of natural taxonomy. Hennig reasoned that the only true classification was one in which every group or taxon could be distinguished by a set of shared, derived characteristics—features shared by other members of the group and with the group's common ancestor and which were absent in organisms from other groups. Such groups—or clades, as Hennig called them—are always monophyletic, meaning that they had just one common ancestor. Hennig's

◄ Tobacco mosaic virus *Scientists classify viruses in various ways: for example, according to their type of genetic material, by the disorders that they cause, or by their size or shape. The structure of the tobacco mosaic virus is relatively basic: a cylindrical capsid of more than 2,000 copies of the same protein surrounding a molecule of RNA.*

system of classification became known as cladistics, and it is recognized as the best way to show the evolutionary relationships of species. However, while most taxonomists accept cladistics as logical, there are difficulties in following it exactly. In a truly cladistic system, levels of classification such as family, order, and class become rather meaningless because in a natural tree of life forks do not occur at conveniently spaced intervals. A clade always includes all the descendants of a common ancestor, not just a subset of them. Thus in a cladistic tree of the vertebrates, birds, and maybe mammals, too, appear as subgroups of reptiles because birds and mammals arose from reptile or reptilelike ancestors. In a traditional phylogenetic tree, reptiles, birds, and mammals are shown as "sister" groups, classes of equal rank.

◄ *Fungi such as this shiitake mushroom* Lentinus edodes *were once thought to be plants. Fungi are now placed in their own kingdom.*

The five-kingdom classification of life

KINGDOM	DESCRIPTION AND DIVISIONS	SPECIES
Monera, or Prokaryota	Single-cell organisms, including bacteria and cyanobacteria.	Many
Protista	17 phyla of single-celled organisms, including Stramenopiles (diatoms and related species) and Sarcodina (amoebas and allies).	Many
Fungi	4 divisions of spore-bearing organisms: Basidiomycota and Ascomycota are the "higher fungi," and Zygomycota and Chitridiomycota are the "lower fungi." Some fungi, such as yeasts, consist of single cells, but more usually fungi consist of a network of tubes called hyphae, bound by cell walls.	About 64,000
Plantae	10 divisions of photosynthetic eukaryotes that develop from embryos and show an alternation between haploid and diploid generations in their life cycles. The largest divisions are the Magnoliophyta (220,000 species, including the flowering plants, or angiosperms) and the Bryophyta (25,000 species of mosses and liverworts).	More than 260,000
Animalia	14 phyla of multicell eukaryotes that develop from a relatively large egg and a small sperm through a series of embyonic stages. The most species-rich phyla are the Arthropoda (1 million species), the Mollusca (110,000 species), and the Chordata (45,000 species).	More than 1,300,000

Another difficulty with cladistics is that it can be very difficult to decide which shared characteristics are truly derived (that is, inherited from an ancestor that was the first organism to possess the trait). Natural selection works so that animals with quite separate ancestries have often evolved similar characteristics in response to similar selective forces. This phenomenon is called convergence because the lineages converge on similar solutions from two different starting points. Convergence explains similarities between organisms such as dolphins and sharks, which have evolved similar adaptations to the challenges of life in the same environment.

Molecular taxonomy

Modern molecular biology has provided 21st-century taxonomists with some powerful new tools. Because we have techniques for sequencing DNA and other large biological molecules such as proteins, we no longer have to rely on appearances, which can be very deceptive. Biologists can now look at the underlying instructions that cause organisms to develop in a certain way. Modern techniques are also revealing new divisions between groups of organisms even greater than those between the old five kingdoms. Most biologists now agree that five kingdoms are nowhere near enough to represent the diversity of life on Earth. The modern tree of life is now drawn with something like 39 kingdoms divided into not two but three supergroups, or domains: the Eucarya, Bacteria, and Archaea.

The concept of domains is credited to the American microbiologist Carl Woese, who defined the Archaea in 1976. Organisms belonging to this group, he claimed, were prokaryotes but not bacteria. Woese's three-domain system is not universally accepted, and many scientists and teachers prefer the five-kingdom system, which places less emphasis on the difficult-to-study prokaryotes. Whichever system becomes generally accepted, one thing seems certain: it will not last. The kingdoms of life, however they are defined, have secrets and surprises yet in store.

AMY-JANE BEER

Evolutionary tree

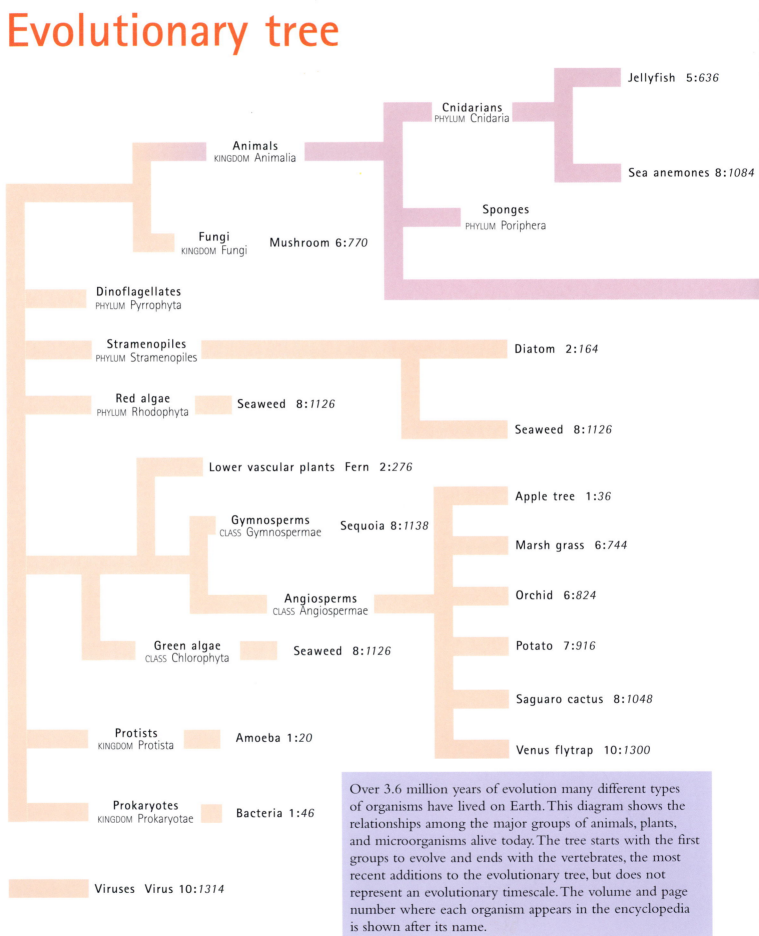

Jellyfish 5:*636*

Cnidarians
PHYLUM Cnidaria

Sea anemones 8:*1084*

Animals
KINGDOM Animalia

Sponges
PHYLUM Poriphera

Fungi
KINGDOM Fungi   Mushroom 6:*770*

Dinoflagellates
PHYLUM Pyrrophyta

Stramenopiles
PHYLUM Stramenopiles   Diatom 2:*164*

Red algae
PHYLUM Rhodophyta   Seaweed 8:*1126*

Seaweed 8:*1126*

Lower vascular plants   Fern 2:*276*

Apple tree 1:*36*

Gymnosperms
CLASS Gymnospermae   Sequoia 8:*1138*

Marsh grass 6:*744*

Orchid 6:*824*

Angiosperms
CLASS Angiospermae

Green algae
CLASS Chlorophyta   Seaweed 8:*1126*

Potato 7:*916*

Saguaro cactus 8:*1048*

Protists
KINGDOM Protista   Amoeba 1:*20*

Venus flytrap 10:*1300*

Over 3.6 million years of evolution many different types of organisms have lived on Earth. This diagram shows the relationships among the major groups of animals, plants, and microorganisms alive today. The tree starts with the first groups to evolve and ends with the vertebrates, the most recent additions to the evolutionary tree, but does not represent an evolutionary timescale. The volume and page number where each organism appears in the encyclopedia is shown after its name.

Prokaryotes
KINGDOM Prokaryotae   Bacteria 1:*46*

Viruses   Virus 10:*1314*

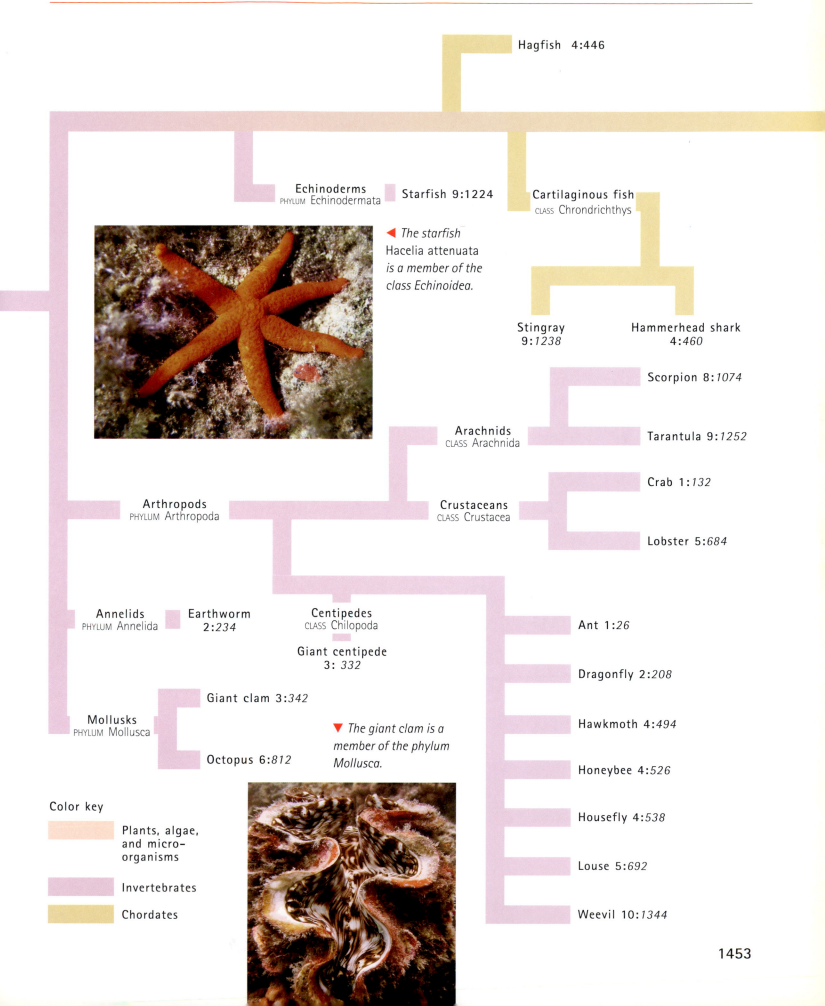

Hagfish 4:446

Echinoderms
PHYLUM Echinodermata

Starfish 9:1224

Cartilaginous fish
CLASS Chrondrichthys

◀ *The starfish*
Hacelia attenuata
is a member of the
class Echinoidea.

Stingray
9:*1238*

Hammerhead shark
4:*460*

Scorpion 8:*1074*

Arachnids
CLASS Arachnida

Tarantula 9:*1252*

Crab 1:*132*

Arthropods
PHYLUM Arthropoda

Crustaceans
CLASS Crustacea

Lobster 5:*684*

Annelids
PHYLUM Annelida

Earthworm
2:*234*

Centipedes
CLASS Chilopoda

Ant 1:*26*

Giant centipede
3: *332*

Dragonfly 2:*208*

Giant clam 3:*342*

Hawkmoth 4:*494*

Mollusks
PHYLUM Mollusca

▼ *The giant clam is a*
member of the phylum
Mollusca.

Honeybee 4:*526*

Octopus 6:*812*

Housefly 4:*538*

Color key

Plants, algae,
and micro-
organisms

Louse 5:*692*

Invertebrates

Chordates

Weevil 10:*1344*

Coelacanth 1:*122*

Lobe-finned fish
SUBCLASS Sarcopterygii

Mammals
CLASS Mammalia

Ray-finned fish
SUBCLASS Actinopterygii

Birds
CLASS Aves

Gulper eel
4:*436*

Sailfish
8:*1060*

Sea horse
8:*1096*

Trout
9:*1280*

Amphibians
CLASS Amphibia

Reptiles
CLASS Reptilia

Bullfrog
1:*54*

Newt
6:*800*

Crocodile
2:*148*

**Green
anaconda**
3:*396*

**Jackson's
chameleon**
5:*624*

**Snapping
turtle**
9:*1194*

Tortoise
9:*1266*

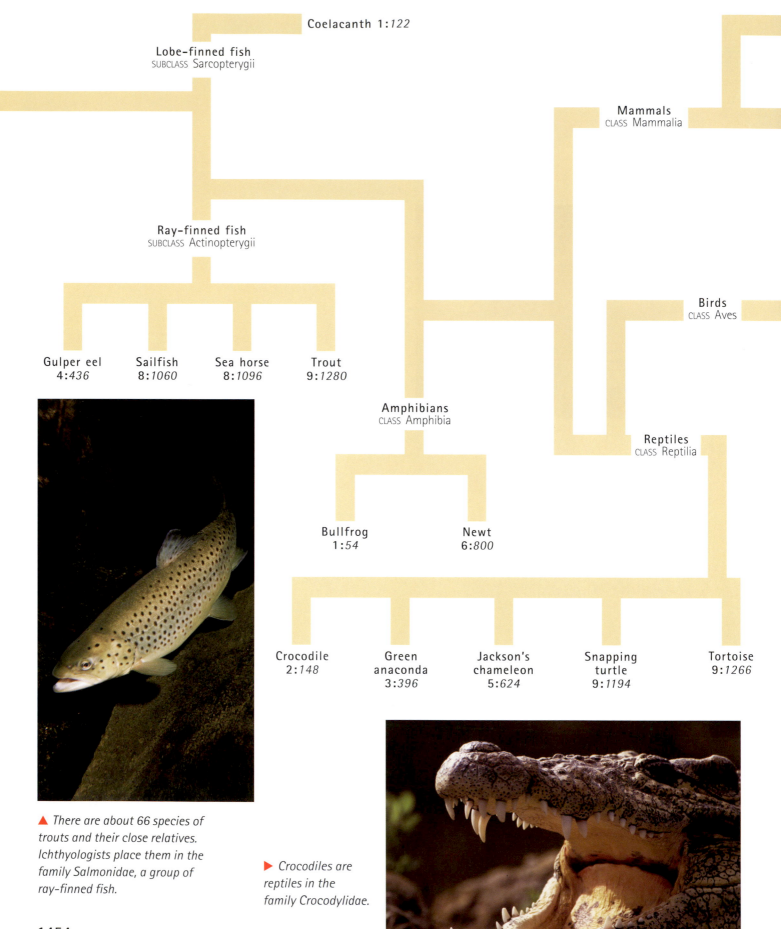

▲ There are about 66 species of
trouts and their close relatives.
Ichthyologists place them in the
family Salmonidae, a group of
ray-finned fish.

► Crocodiles are
reptiles in the
family Crocodylidae.

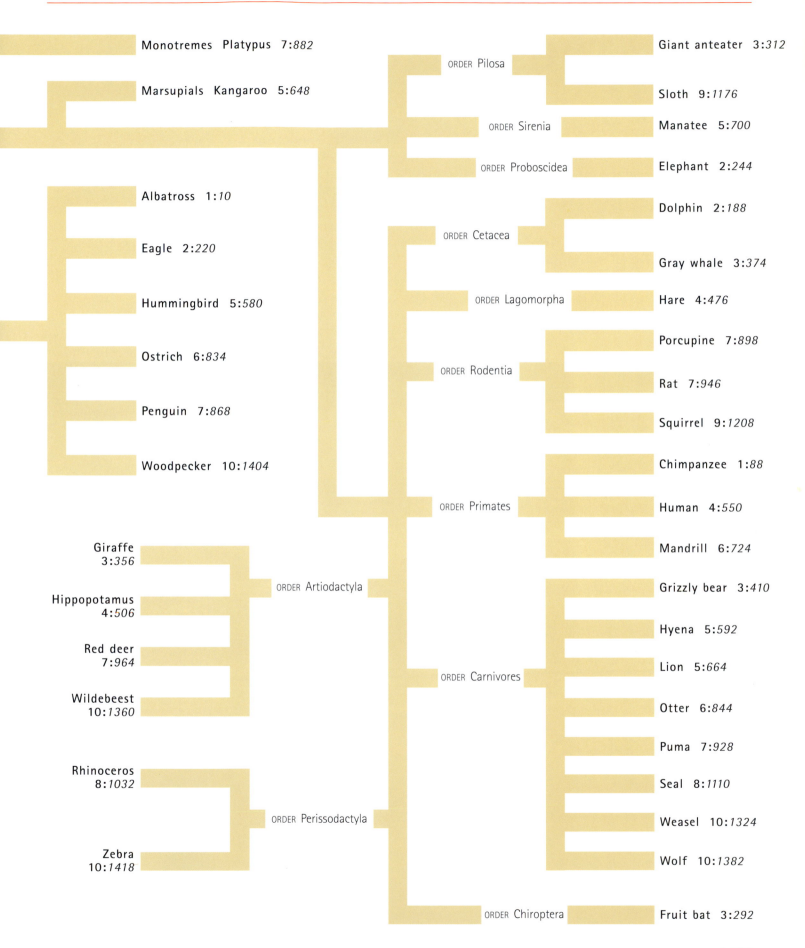

The study of anatomy

People have been studying anatomy in its loosest sense for tens of thousands of years. The bodies of animals, including humans, have always been of interest, although to early humans this interest was far from academic or scientific. Early knowledge of anatomy was a matter of survival. For early humans it was important to know which parts of another animal were good to eat and which parts could be used for other purposes—bone, horn, hide, and fur all had important uses. A basic appreciation of anatomy also helped make early humans more effective predators: hunters soon learned that a spear thrust into the heart would bring down an animal faster than one in the rump; that breaking the spinal cord or cutting the blood vessels of the neck would kill almost instantly; and that a heavy blow to the head was more damaging than one elsewhere. Through accidental and deliberate injury, these same grisly lessons must have been learned about our own species, too.

The earliest records of the formal study of anatomy come from ancient Greece. Hippocrates (c. 460–c. 380 BCE), known as the father of medicine, was among the first to

▼ The ancient Greek philosopher Aristotle was one of the first people to recognize that dolphins are mammals and not fish.

Cave art

In some parts of the prehistoric world it is clear that appreciation of basic animal form was well advanced tens of thousands of years ago. Cave art such as the famous 20,000-year-old paintings at Lascaux in France show animals that are clearly identifiable as horses, bison, aurochs (wild cattle), and mammoths. In 1994 equally striking images were discovered at the Chauvet caves, also in France. The drawings caused even more excitement when carbon dating revealed them to be 35,000 years old.

▶ *Cave paintings, sometimes thousands of years old, show the importance animals have always played in people's lives.*

attempt an anatomical study of the human body. However, in Greece, as in most cultures, the dissection of humans was considered an abomination: the dead were not to be violated. As a result, much of what was written or taught regarding the anatomy of humans was in fact taken from the study of animals such as cattle and monkeys. The study of human and animal anatomy thus went hand in hand, and it was at least 2,000 years before the two were clearly set apart. The first written account on the subject of anatomy is probably a document entitled *On the Heart*, written in the fifth century BCE by Hippocrates or one of his students. It compares the human heart with that of other mammals.

The Greek naturalist and philosopher Aristotle (384–322 BCE) took an active interest in all aspects of natural history and is generally considered the father of anatomy and of biology in general. He drew on his own observations and the knowledge of others, including hunters and fishers, to write extensively on the external and internal anatomy of animals in his books *History of Animals*, *Parts of Animals*, and *Generation of Animals*. Aristotle was the first to describe, among many other things, the four-chamber stomach of ruminants, the uterus of female mammals and birds, and the embryonic development of chicks and dogfish. He also wrote on the anatomy of humans, describing the external features in great detail but making clumsy errors regarding internal structures, suggesting for example that a man had only eight ribs. These mistakes make it clear that Aristotle never actually dissected a human and was bound by the strict laws, traditions, and natural squeamishness associated with such activities. The first person to seriously study human anatomy was another Greek, Herophilus (335–280 BCE), who along with

The age of microscopy

The microscope was the advance anatomists needed to take their studies to a new level of detail. The Englishman Robert Hooke used an early microscope to examine plant tissues and he is credited with the "discovery" of cells in 1667. The Dutchman Antoni van Leeuwenhoek made simple microscopes in which the magnification was provided by a single lens made from a drop of glass. He used these simple devices with such skill he was able to describe the first known protists in the late 17th century and early 18th century and to dissect such tiny animals as fleas, aphids, mites, and ants.

his contemporary Erasistratus (c. 304–c. 250 BCE), is said to have performed dissections on living men as well as dead bodies. The unfortunate live subjects would have been prisoners sentenced to a terrible punishment for their crimes.

The last of the great ancient Greek anatomists and perhaps the most important figure in the ancient history of the subject was Claudius Galen (129–216 CE). However, like Aristotle 500 years before, Galen made mistakes when it came to humans because his dissection subjects were other animals: dogs, cattle, and monkeys. Galen's writings were later translated by scholars who did not think to question them, and they became the standard references for physicians of the Middle Ages. It was centuries before many of Galen's errors began to be corrected.

The new anatomy

During the Renaissance in Europe there were notable advances in the study of human anatomy, not least by the artist and scientist Leonardo da Vinci (1452–1519), whose drawings were as detailed as any produced since, and whose study of the human muscular system is still considered one of the greatest anatomical works of all time.

However, it was the Belgian Andreas Vesalius (1514–1564) who finally broke with

tradition and put right some of the errors made over 1,300 years before by Galen. In preparing his book *De Humani Corporis Fabrica* (*The Construction of the Human Body*), Vesalius diligently performed hundreds of human dissections on people. Thus he was able to demonstrate than many of Galen's descriptions actually referred to monkeys and apes, not to humans.

Until the time of Vesalius's work, studies of anatomy had been concerned mainly with structure, rather than function. Anatomists simply described what was there. Some made assumptions about how various body parts might work, but these were very rarely backed up by any scientific proof. For example, it was known that the heart pumped blood, but until 1628, when the English physician William Harvey (1578–1657) described the network of veins and arteries in a variety of animals and in humans and demonstrated one-way circulation through them, it was always assumed that blood somehow flowed in both directions within the same vessels. Harvey was unable to explain exactly how blood passed from arteries to veins, but this problem was solved a few years

◀ *Charles Darwin was able to explain that evolution was responsible for homology—shared anatomical features among different life-forms.*

▶ *Blood circulates around the body of vertebrates (this is an American bullfrog). Oxygen-rich blood from the lungs is pumped in arteries (red on the diagram), and oxygen-poor blood returns in veins (blue). Until William Harvey's work on the circulatory system in the 17th century, people believed blood flowed in both directions in the same vessels.*

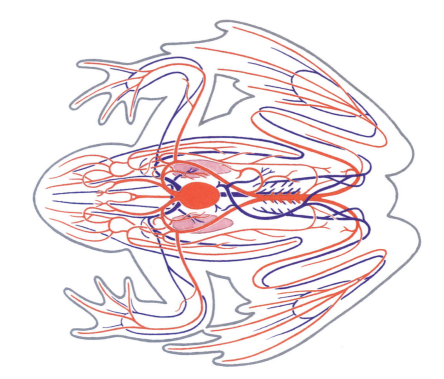

later by the Italian physiologist Marcello Malpighi (1628–1694). With the aid of the newly invented compound microscope, Malpighi was able to observe minute vessels, or capillaries, in the tissues of frogs. Malpighi was a keen advocate of comparative anatomy. He engaged in the study of animals for its own sake, not just for what it could reveal about the human body. Also, he was the first anatomist to publish detailed information on the anatomy of an invertebrate animal, the silk moth *Bombyx mori*.

In the mid-18th century two Scottish brothers, William and John Hunter (1718–1783 and 1728–1793, respectively), made their names as great anatomists. William established a celebrated school of anatomy in London, and while he taught, his younger brother John devoted himself to research, performing dissections on all parts of the human body and on hundreds of different animal species. John Hunter was the first person to attempt a scientific study of teeth, naming incisors, molars, cuspids, and bicuspids.

Anatomy comes of age

As anatomists became more skilled, they began to grasp an important new concept, which became known as homology, meaning

IN FOCUS

Body snatchers

In the early 19th century human anatomists were always short of subject material. The law granted them a number of bodies (usually those of executed criminals), but these were never enough. Thus began the unsavory practice of body-snatching. Students of anatomy sometimes raided fresh graves and stole the bodies for dissection. The most famous body snatchers, William Burke and William Hare, went even further. They murdered 16 people in Edinburgh, Scotland, in 1827 and 1828 and sold the bodies to the surgeon Robert Knox (1791–1862), who chose to turn a blind eye to where his subjects came from. Burke and Hare were eventually caught, and Burke was convicted and hanged. His body was publicly dissected, and his skeleton was placed on display in the University Medical School, Edinburgh, where it remains. Hare went free and Knox was never prosecuted, but both were forced to flee Edinburgh and ultimately died in poverty.

IN FOCUS

The visible human project

In 1994 the United States National Library of Medicine published the first results of its Visible Human Project online. Researchers used two techniques called computed tomography and magnetic resonance on a series of 0.04-inch (1-mm) cryosections (taken from a frozen body so that the samples would hold their shape) of a man's body to create a unique interactive resource. Computed tomography uses X-ray scans from many different angles to produce a 3-D image, and magnetic resonance uses high-frequency radio waves to give a detailed image of the body.

This set of data allows students or anyone else interested in human anatomy to see inside the human body. The following year, data for a woman were added—this time in even greater detail, with each cryosection just 0.01 inch (0.3 mm) thick. Anyone can visit the Visible Human Project website at: **www.nlm.nih.gov/research/ visible/visible_human.html**

look very different but are homologous because at a fundamental level they have important similarities: they share much the same arrangement of bones. We now know that this is because they all evolved from the forelimb of the same common ancestor, but before Charles Darwin's *Origin of Species,* anatomists could record these interesting facts but had no satisfactory explanation for them.

The opposite of homology is homoplasy, the study of structures that look superficially the same (such as the flipper of a dolphin and the fin of a fish) but have quite different origins. The study of anatomy revealed a great many such features, but as the Russian-American biologist Theodosius Dobzhansky (1900–1975) later observed, "Nothing in biology makes sense except in the light of evolution." By the time Darwin came to set out his theory of evolution by means of natural selection, much of the evidence he needed was readily available in the collected specimens and drawings of generations of dedicated anatomists.

Modern anatomical techniques

For more than 2,000 years the anatomist's tools were simple and changed little: blades, forceps, needles, and pointers. The invention of the light microscope heralded a new era, but the modern anatomist is now able to examine his or her subject in detail without touching a knife. With the aid of modern imaging technology, including X–rays, endoscopes,

▼ *This X-ray image shows a replacement knee, including a metal plate and screw (left), and a normal knee. X-ray technology has made the study of skeletal anatomy much easier.*

"sameness." Homology describes shared features of animals that are similar in a fundamental way. The wing of a bird, the flippers of a dolphin, and the arm of an ape

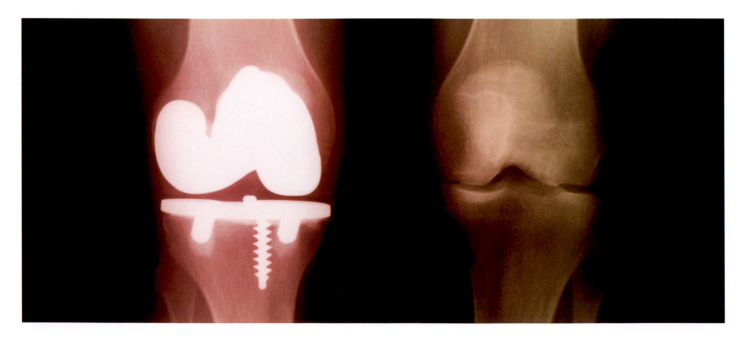

magnetic resonance imaging (MRI), computed tomography (CT), miniature cameras, and ultrasound, the internal structures of humans and animals can be examined in detail while the organism is alive and fully intact. Not only can these techniques allow anatomists to see inside living bodies; advanced computer simulations allow them to examine structures in three dimensions, see how they function (for example, how the ribs move when the diaphragm muscle contracts and expands), and perform virtual dissections.

Cellular anatomy and ultrastructure

Improvements in microscope technology have continued since the days of Leeuwenhoek and Hooke. As a result of improvements in lens technology, light microscopes are now routinely used to achieve magnifications of 1000 times. Anatomy at the cellular level has revealed some of the fundamental differences between major groups of organisms: for example, the fact that bacterial cells lack membrane-bound organelles; and the way that animal epithelial cells, such as the lining of the gut or the epidermis, separate the insides of the animal from the outside but connect to one another with special junctions that allow chemical signals to pass from cell to cell. In the 1930s and 1940s microscopes were developed that used beams of electrons instead of light. Modern electron microscopes can achieve magnifications of hundreds of thousands of times, allowing microbiologists to describe in very fine detail the anatomy of cells and their contents. There are two kinds of electron microscope: transmission EM shows the detail in ultrathin sections of tissue; scanning EM is used to examine surface detail.

Anatomy: Applications and careers

The applications of modern anatomy are many and varied. Anatomy in some form is usually taught as part of degree courses in biology, physiology, medicine, and art. The relevance to medical science is obvious, since an understanding of pathology, physiotherapy, and prosthetics requires detailed knowledge of the affected body parts. The growing study of sports science also depends on an understanding of human anatomy. In broader

▲ *A scientist uses a scanning electron microscope (SEM). This technology has made possible the study of cell structure and microorganisms.*

zoology, anatomical knowledge is used by taxonomists, physiologists, and paleontologists, among others. More recently anatomists have sometimes been called on to lend their expertise in bringing convincing computer-generated effects to the cinema screen and games consoles.

AMY-JANE BEER

Glossary

Abductor
A muscle that moves a limb away from the body of an animal. Abductor muscles work antagonistically with adductor muscles.

Acid
Any substance that gives up hydrogen ions (H^+) in solution, increasing the overall hydrogen ion concentration of the solution.

Actin
Protein that is present in microfilaments, such as those that enable muscles to contract.

Action potential
A change in the voltage across the membrane of a nerve cell when an impulse passes along it. This results from a change in the permeability of the membrane, which causes sodium ions to pass into the cell and potassium ions to move out.

Adaptation
Accumulation of inherited characteristics or a certain genetically based characteristic or behavior that makes an organism suited to its environment and way of life.

Adaptive radiation
Evolutionary diversification of a single species, with production of many different species adapted to different environments.

Adductor
A muscle that moves a limb toward the body of an animal. Adductor muscles work antagonistically with abductors.

Adenine
Nitrogen-containing base that is present in nucleic acids such as DNA and RNA.

Adrenal gland
Endocrine gland located close to the kidneys, consisting of an inner medulla and an outer cortex that produce steroids and hormones.

Aerobic
Describes a process that takes place, or an organism that grows or metabolizes, in the presence of gaseous or dissolved oxygen.

Algae
Any of a group of simple organisms that, like plants, can make their own food. Unlike plants, algae lack roots, leaves, and flowers. Seaweeds are algae.

Alimentary canal
A roughly tubular organ where food is ingested, digested, and absorbed. In most animals there are two openings—the mouth and the anus. The esophagus, stomach, small intestine, and large intestine are all component sections.

Alkali
Soluble base or a solution made up from a base.

Alternation of generations
The occurrence within the life cycle of an organism of two of more forms, which differ in appearance, habits, and method of reproduction. For example, in plants the sexual form is called the gametophyte, and the asexual generation is the sporophyte.

Alveolus
One of the thousands of tiny air sacs at the end of each bronchiole in the lungs of mammals and reptiles. Alveoli increase the surface area available for gas exchange.

Ampullary organ
A type of sense organ called an electroreceptor in some fish, such as rays, and certain amphibians, such as sirens, that detects the electricity produced by the muscles of living animals, giving a means of detecting prey or predators. In sharks this organ is called the ampulla of Lorenzini.

Anaerobic
Describes a process that takes place, or an organism that grows or metabolizes, in the absence of gaseous or dissolved oxygen.

Angiosperm
A flowering plant usually with seeds that are enclosed in carpels, or fruits. The other main group of plants is the gymnosperms.

Anther
The male part of a flower that produces pollen.

Antibody
Protein compound produced by vertebrate plasma cells

that binds to foreign bodies (antigens), which then clump together and can be destroyed by white blood cells.

Antigen
Any molecule that can stimulate an immune response, inducing the production of a specific antibody.

Aorta
Largest artery of the mammalian body, leaving the heart from the left ventricle and supplying all parts of the body with oxygenated blood.

Arboreal
A term that describes animals that live in trees.

Artery
A vessel that carries blood from the heart to the rest of the body. Most arteries carry oxygen-rich blood, but the pulmonary arteries carry oxygen-poor blood from the heart to the lungs.

Arthropod
An animal with a hard outer covering (exoskeleton) and several pairs of jointed limbs. Arthropods include insects, arachnids, and crustaceans.

Asexual reproduction
Reproduction by an individual organism that occurs without mating or mixing genes with another individual, producing clones; often occurs through dividing in two, as in some starfish and sea anemones, and many plants.

Axon
Extension of neuron that transmits nerve impulses away from the cell body.

Bacteria
A phylum of diverse single-cell organisms that lack a cell membrane but have a distinctive cell wall.

Bilateral symmetry
An arrangement in which one half of an organism is a mirror image of the other half if an imaginary line is drawn along the longest axis.

Binocular vision
Vision through two eyes pointed in the same direction. It restricts the area of view but allows accurate judging of distance. Many predators have binocular vision. When the eyes are positioned on the side of the head they give nonbinocular vision.

Biodiversity
Diversity of plant and animal species in an environment.

Bioluminescence
The production of light by an organism. Some species of plankton, fish, and insects are bioluminescent.

Biomass
Total quantity of organic matter in a region or habitat.

Bipedal
Walking on two limbs. Humans and birds are bipedal, but most mammals, reptiles, and amphibians are quadrupedal.

Bladder
An elastic-walled organ in the urinary systems of some reptiles and fish, most amphibians, and all mammals that acts as a receptacle for urine before excretion.

Blood-brain barrier
The protective membrane that controls the passage of substances from the blood to the fluid bathing the brain and spinal cord.

Book lung
A series of leaflike plates (lamellae) through which scorpions and some spiders breathe.

Branchial
Relating to the gills of a fish or amphibian.

Bronchi
The tube that branches from the trachea and passes to the lung in vertebrates.

Brood
To sit on or incubate eggs or live young. Also, a group of young of the same age.

Buccal cavity
The mouth cavity.

Camouflage
Coloring, shape, or texture of the exterior of an organism that makes it hard to see. A camouflaged animal may match its surroundings (cryptic camouflage) or have patterns that break up its outline (disruptive camouflage).

Canine tooth
A sharp, pointed tooth found mainly in carnivores that is suited to tearing meat. There are two canines in each jaw,

situated between the incisors and premolars. Some herbivorous mammals do not have canines.

Capillary

The narrowest type of blood vessel. Capillary walls consist of a single layer of cells, so nutrients, dissolved gases, and waste products can pass through them.

Carapace

The domed upper part of the shell of tortoises and turtles; the upper surface of the exoskeleton of some crustaceans, such as crabs.

Carbohydrate

Compound that contains carbon, hydrogen, and oxygen in the ratio 1:2:1. Most are produced by green plants and provide and important food source for animals.

Carbon dioxide

A gas formed as a by-product of respiration in animals and plants, and when fuel such as oil is burned. This gas is thought to be a main cause of the greenhouse effect, a process by which energy from sunlight is trapped within Earth's atmosphere.

Cardiac muscle

A type of muscle that is unique to the vertebrate heart.

Carnassial teeth

A powerful scissorlike pair of teeth possessed by most carnivores apart from seals.

Carnivore

An animal or plant that feeds on animals. Also, members of the order Carnivora, such as bears, cats, and dogs.

Carpel

A structure in flowering plants that encloses the ovules; it comprises the ovary, stigma, and style.

Cartilage

A tough, elastic, gristly, and somewhat translucent tissue that holds together the skeleton of most vertebrates. In sharks and rays, the skeleton is made of cartilage.

Cecum

Blind-ending structure present in the digestive system of some animals; it may house cellulose-digesting bacteria.

Cell

The basic structural and functional unit of all living organisms except viruses. Organisms may consist of a single cell or—as with complex vertebrates— billions of cells.

Cellulose

A tough substance found in plant cell walls. It cannot be digested in most animals without the aid of bacteria in the stomach or intestine.

Central nervous system

That part of the nervous system that coordinates nerve functions. In vertebrates it is made up of the brain and spinal cord, and in invertebrates it comprises nerve cords and ganglia.

Cephalothorax

The fused head and thorax of a spider.

Cervical vertebrae

The individual bones (vertebrae) of the neck.

Chelicerae

Appendages near an arachnid's mouth; those of spiders carry fangs.

Chemical defense

A chemical that is secreted by an animal to protect it from a predator. The chemical may be foul-smelling or foul-tasting, so repelling the predator; the chemical may disguise the animal's own smell, thus confusing the predator; or it may be a toxin that poisons the predator.

Chemoreceptor

A sense organ, such as a taste bud, that responds to chemical stimuli.

Chitin

A tough material in the exoskeleton of arthropods, such as insects, and in the cell walls of many fungi.

Chlorenchyma

Photosynthetic plant cells that contain chloroplasts.

Chlorophyll

The green pigment found in plant and algal cells that traps solar energy during the process of photosynthesis.

Chloroplast

An organelle (mini-organ) inside plant cells which contains chlorophyll and in which photosynthesis takes place.

Chromatophore

A cell that contains pigment in the skin of some vertebrates, such as chameleons. Dispersion of the pigment in the cells can cause the skin to change color.

Chromosome

Long strand of coiled DNA composed of genes, the genetic information for most organisms.

Chrysalis

The pupa of a butterfly or moth.

Cilium

A tiny, beating hairlike structure on the surface of cells. Many cilia beating together can produce a current or move a single cell.

Cladistics

A method of classifying organisms in which animals and plants are placed in taxonomic groups, or clades, strictly according to their evolutionary relationships.

Cloaca

Chamber into which lead the urinary, intestinal, and reproductive canals of birds, reptiles, amphibians, and some mammals.

Clone

An organism produced by asexual reproduction, which is, therefore, an exact genetic copy of its parent.

Coelom

The fluid-filled body cavity of vertebrate and most invertebrate animals.

Colloid

Substance, such as gelatin or starch, that will not pass through a membrane.

Compound eye

Eye that consists of many smaller, light-sensitive units called ommatidia; often present in insects.

Conifer

A cone-bearing plant, such as a pine tree. Most conifers have needle-shaped or conelike leaves, and many retain leaves throughout the year.

Convergent evolution

Similar behavior or appearance of two organisms with similar lifestyles that are not closely related.

Countercurrent heat exchange

A mechanism by which some of an animal's heat is transferred back into the body. In a penguin, for example, the arrangement of blood vessels ensures that cooler blood returning from the feet is warmed by adjacent vessels carrying warmer blood from the bird's core.

Coxa

The first segment of the leg of an insect, an arachnid, or certain other arthropods.

Crassulacean acid metabolism

A type of photosynthesis that allows plants living in arid environments to keep their stomata closed during the day and thus reduce water loss.

Cuticle

A waxy layer on the surface of plants; the horny noncellular exoskeleton that is secreted by epidermal cells in many invertebrates.

Cytoplasm

The material surrounding the nucleus of a cell, including the inner endoplasm and the outer ectoplasm.

Denticle

A toothlike scale, also called a placoid scale, found on sharks and rays. Denticles aid streamlining and strengthen and toughen the skin.

Dicotyledon

A flowering plant in the class Dicotyledonae whose germinating seeds produce two seed leaves. These plants are also called dicots.

Diffusion

The movement of particles, such as a gas, from a region of high concentration to a region of lower concentration.

Digestion

The process of breaking down food into easily absorbed substances.

Digitigrade

The gait of most fast-running mammals, in which only the toes make contact with the ground, and the rest of the foot is raised off the ground.

Diurnal

A term meaning active during the day; the opposite of nocturnal, active during the night.

Divergent evolution

Evolution over time of a number of different species from one species of living organism.

DNA (deoxyribonucleic acid)

A molecule inside cells that carries genetic information and directs many of the cell's activities.

Domain

A taxonomic category above the level of kingdom in some classifications of organisms. The three domains are Archaea, Bacteria, and Eukarya.

Dormancy

A condition in which an animal or plant's metabolism slows down; often used by organisms to better survive periods of harsh environmental conditions.

Echolocation

Use of sound waves to find the way and detect prey in the dark. Bats, dolphins, and some birds use echolocation.

Ecosystem

Community of organisms and their interactions with one another, together with the environment in which they live and with which they interact.

Ectotherm

An animal which cannot regulate its body temperature internally and whose body temperature, therefore, depends on the environment; also called "cold-blooded." Fish, amphibians, reptiles, and invertebrates are ectotherms, whereas birds and mammals are endotherms.

Electroplates

Electricity-producing cells in fish such as electric rays and electric eels. The electric current kills or stuns prey or deters predators.

Electroreceptor

A type of organ in many fish, such as rays and sharks, that detects electric fields created by the muscle action of animals; used to track prey in dark water.

Elytron

One of a pair of wing cases that protect the hind (flight) wings in beetles.

Embryo

The first stage of an animal's or a plant's life cycle after fertilization of the egg.

Enzyme

Protein that catalyses chemical reactions in organisms.

Epidermis

The outer layer of cells that covers the body of plants and animals.

Esophagus

The muscular tube by which food travels from the mouth to the stomach in vertebrates and many invertebrates.

Estrus

Period of sexual receptivity occurring in female mammals around the time of ovulation.

Eukaryotic cell

Cell in which the chromosomal genetic material is contained within one or more nuclei and is separated from the cytoplasm by two nuclear membranes.

Evolution

Any cumulative genetic change that occurs in a population of organisms from one generation to the next. Over many generations, evolution can change the structure of an animal or plant.

Exoskeleton

A hard outer covering of certain animals, such as the chitinous cuticle of insects, the bony plates of armadillos, and the shell of tortoises and some mollusks.

Gene

Discrete unit of hereditary information present in a chromosome.

Genus

Taxonomic grouping of very similar organisms thought to be closely related species.

Gill

A breathing organ used by many types of aquatic animals to gain oxygen. Fish, young amphibians, and many aquatic invertebrates have gills. The term is also used to describe the spore-producing ridges of tissue radiating from the center of the underside of a mushroom cap.

Graviportal

A type of skeletal structure in which the limbs are straight and very sturdy and bear a large body mass. An elephant has a graviportal skeleton in contrast with a rhinoceros's mediportal structure and a cheetah's cursorial structure.

Gymnosperm
A seed-bearing plant that, unlike an angiosperm, has naked seeds, often carried in cones.

Hallux
Innermost digit on the hind limb of a vertebrate. For example, it is the big toe in humans.

Haploid
A type of cell containing only one copy of each chromosomes. Gametes are haploid cells.

Hemocoel
A cavity or series of spaces between the organs of most arthropods and mollusks through which bloodlike hemolymph moves.

Hemocyanin
A copper-containing blue pigment found in the hemolymph (bloodlike fluid) of spiders, crabs, lobsters, and some mollusks, which transports oxygen.

Hemoglobin
An iron-containing red pigment that transports oxygen; found in vertebrate red blood cells and in some invertebrates.

Hemolymph
A fluid similar to vertebrate blood pumped around the arthropod's body by its heart.

Herbivore
An animal that eats only plants.

Hermaphrodite
An animal that has both male and female reproductive organs. For example, earthworms are hermaphrodites.

Heterotrophs
Organisms that cannot make their food from inorganic substances and feed at the expense of other organisms or on decaying matter.

Hexapod
Any animal that has six legs, such as an insect.

Hormone
A chemical messenger transported around the body by the blood or hemolymph. Hormones are secreted in small quantities by glands or nerve cells and can affect tissues in a distant part of the body.

Hyoid bone
A muscle and bone structure inside the throat of some birds and lizards; used to extend the tongue and swallow prey.

Invertebrate
An animal without a backbone (spine).

Jacobson's organ
Either one of a pair of small pits or sacs situated in the roof of the mouth and developed as chemoreceptors in amphibians, reptiles, and some mammals.

Larva
The young form of many invertebrates, fish, and amphibians that looks very different from the adult, lives in a different habitat, and eats different foods.

Larynx
Voice box containing vocal cords. The larynx is situated at the anterior end of the trachea.

Lateral line
A canal running along the side of the body of some animals that enables them to detect prey and predators. The canal contains pores, which open internally into tubes. The tubes have sensory organs.

Leukocyte
White blood cell.

Ligament
Vertebrate connective tissue that joins bones.

Lymph
Clear fluid consisting mostly of water and dissolved salts and proteins that flows in vessels of the lymphatic system.

Lymphocyte
White blood cell that is responsible for immune responses.

Maggot
The larva of a fly.

Malpighian tubule
Long, slender tube that is part of the excretory system in many arthropods.

Marsupium
The abdominal pouch of marsupial mammals, such as kangaroos and koala bears, in which young are reared.

Mediportal

A type of skeleton with many adaptations for bearing heavy loads, including thick limb bones and thick ankle and wrist bones. A hippopotamus has a mediportal skeleton.

Medusa

The free-swimming phase in the life cycle of cnidarians such as jellyfish; the polyp stage is the sedentary phase.

Meiosis

A form of nuclear division in which, after chromosome duplication in a reproductive cell, the diploid parent nucleus divides twice, forming four haploid offspring cells.

Metabolism

The chemical changes in living cells by which energy is produced for vital processes in the body.

Metamorphosis

The change from young form to adult form in animals with a larval stage.

Microorganism

Any living organism that is invisible to the naked eye but is visible under a microscope. Microorganisms include viruses, bacteria, and some algae.

Microvilli

Protuberances in the wall of the gut of an arthropod that increase its surface area.

Mitosis

A type of nuclear division that results in two offspring cells, each having a nucleus containing the same number and kind of chromosomes as the parent cell.

Molars

Large teeth at the back of the mammalian mouth.

Molt

The shedding of the exoskeleton by an arthropod or the skin of a reptile as it grows; the seasonal loss of feathers in birds, or of fur or hair in mammals.

Monocotyledon

A flowering plant whose germinating seeds produce on seed leaf; also called a monocot.

Mucus

A viscous, slimy fluid that is produced by, and protects, mucous membranes.

Mutualism

A relationship between unrelated organisms from which each benefits.

Myoglobin

A protein of vertebrate muscle fibers that binds to molecular oxygen.

Natural selection

Most widely accepted theory concerning the main mechanism of evolutionary change. The genetic composition of evolutionary lineage changes over time by a nonrandom transmission of genes from one parental generation to the next. Selection of gene combinations will favor those that are best suited to a particular environment.

Nematocyst

Structure present in the stinging cells of cnidarians such as jellyfish and used for defense and the capture of prey.

Neotony

Slowing of the rate of growth and the development of certain body parts relative to the reproductive organs; a neotonous organism reaches sexual maturity in a larval or another immature stage.

Neuron

An elongated cell forming part of the nervous system through which electrical and chemical signals pass around the body. When neurons are collected in large numbers—for example, to form a brain—they process as well as transfer information.

Nocturnal

An organism that is active at night; the opposite of diurnal. Most bats and owls, and many invertebrates are nocturnal, for example.

Notochord

Flexible, longitudinal rod that acts as an internal skeleton in the embryos of all chordates and is retained in the adults of some.

Nucleus

The central, membrane-enclosed part of a cell, containing the chromosomes and acting as the cell's control center.

Nutrient

Any material taken in by a living organism that allows it to grow or replace lost or damaged tissue, and provides energy for metabolism.

Nymph
The young stage of certain insects, such as grasshoppers, earwigs, and dragonflies. The nymph resembles the adult but does not have wings or functional reproductive organs.

Ocellus
A simple eye occurring in insects and other invertebrates; for example, moths have a pair of ocelli. Also, an eyelike marking on the wings of some butterflies and moths.

Ommatidium
One of the numerous units that make up the compound eye of insects and crustaceans.

Omnivore
An animals that feed on both animals and plants.

Opposable
Capable of being placed opposite and against another digit. Humans, for example, have opposable thumbs.

Ossicles
The small bones in the middle ear of vertebrates. Also, the skin plates of many echinoderms.

Osteoderm
Bony plates under a reptile's scales that armor-plate the skin. Crocodilians and reptiles have osteoderms.

Ovary
The organ in female animals in which eggs are produced. In most vertebrates there are two ovaries.

Ovipary
A method of reproduction in which the fertilized eggs are laid or spawned by the mother and hatch outside her body.

Ovipositor
An egg-laying tube found in female insects and some other animals, such as female sea horses.

Ovule
The female gamete, or sex cell, of a seed plant. It has a small opening called a micropyle through which pollen grains enter to fertilize the ovule. The fertilized ovule develops into a seed.

Parasite
An organism that feeds on another living organism, or host. The host may be damaged but is not killed by the parasite.

Parthenogenesis
The production of offspring from an unfertilized egg. It can occur in many plants and a few animals.

Pathogen
Disease-causing organism.

Pedipalps
A pair of appendages near the mouthparts of spiders, used for tasting and feeling; male spiders also use them to transfer sperm to females.

Peristalsis
Contractions of muscle that occur in the walls of hollow organs, such as parts of the digestive tract, that move the contents of the organ through the tube.

Petiole
The narrow waist between the thorax and abdomen of ants and wasps; a leaf stalk in plants.

Pheromone
A chemical released by an animal, often to attract mates.

Phloem
Vascular tissue that transports nutrients, such as amino acids and sugars, in plants.

Photosynthesis
The chemical process by which green plants synthesize organic compounds from carbon dioxide and water in the presence of sunlight.

Placenta
The temporary organ that forms inside a female animal that nourishes the young.

Plantigrade
The gait of many mammals, including humans, in which the whole lower surface of the foot is on the ground.

Plasma
The fluid part of blood, excluding the blood cells. It consists of water and many dissolved substances, including salts, proteins, fats, amino acids, hormones, vitamins, and excretory materials.

Plastron
The tough, flattish plate on the underside of a turtle's shell. Also, a thin layer of air around an aquatic invertebrate's body that is used as a source of oxygen.

Pollination
The transfer of pollen (male sex cells) from one flower to another, either by the wind or by animals such as insects, allowing seeds to form.

Pollex
Innermost digit on the forelimb of a vertebrate. In humans, for example, the thumb. May be absent in some vertebrates.

Polyp
The sedentary phase in the life cycle of some cnidarians; the other life-form is the free-swimming medusa stage.

Proboscis
The tubelike mouthpart of moths, butterflies, and some flies, through which liquids are sucked; the long flexible "nose" of an elephant, rhinoceros, or tapir.

Pupa
The stage during which the larva of certain insects, such as flies and butterflies, transform into an adult. During pupation, the insect does not feed or move around.

Receptor
A cell or group of cells that detects specific stimuli such as heat and pressure.

Rhizome
An underground plant stem that generally bears leaf scars from which new plants can grow. If a rhizome is cut it does not die like a root but develops into more than one plant.

Ribosome
A particle within a cell that acts as the site of protein synthesis. Ribosomes "translate" messenger RNA (mRNA) into protein by using its chemically coded instructions to link amino acids in a specific order and thus make a strand of a particular protein.

RNA
Ribonucleic acid, an organic compound in living cells that is concerned with protein synthesis.

Rostrum
A piercing beak used by insects and arachnids to suck juices from plants or animals.

Ruminant
A hoofed herbivorous mammal, such as a deer or a goat, that chews cud. Ruminants have a multichamber stomach.

Scent marking
A behavior in which some animals leaving strong-smelling secretions on the ground or vegetation to warn off rivals.

Sensor
A receptor cell or group of cells that reacts to a stimulus such as light or the presence of certain chemicals.

Septum
A dividing wall such as that between the different chambers of a heart.

Sexual dimorphism
Difference in appearance of males and females in a species—for example, difference in color or size.

Speciation
The development of a new type of species from an existing species. It occurs when different populations diverge so much from the parent populations that interbreeding can no longer take place between them.

Sperm
A male sex cell that can fuse with a female egg cell to form a new individual.

Spermatophore
A packet of sperm that the males of some amphibians, arachnids, and cephalopods deposit near or within females to achieve internal fertilization.

Spinneret
The silk-spinning organ at the rear of a spider's abdomen.

Spiracle
An opening in the exoskeleton through which insects and some other terrestrial arthropods breathe.

Spore
A single-cell reproductive unit produced by ferns and certain other plants, fungi, and bacteria.

Stigma
Part of a flower that receives pollen grains. The pollen grain grows through the stigma to reach the ovary.

Style
The stalk of carpel, between the stigma and the ovary; in many plants it is elongated to aid pollination.

Symbiosis
A biological relationship between two species.

Synapse
The junction between two nerve cells or between a nerve cell and a muscle; the latter is also called a neuromuscular junction.

Telson
The structure at the end of a scorpion's tail that contains the stinger.

Tendon
Connective tissue that joins two muscles together or joins a muscle to bone.

Testosterone
Vertebrate steroid male sex hormone that is produced by the testes.

Thermoregulation
The general mechanism by which a life-form controls its body temperature. In ectotherms the body temperature depends on the temperature of the environment. Endotherms (mammals and birds) have internal mechanisms for maintaining their body temperature at a level that is usually warmer than their environment.

Thorax
The body region of a vertebrate containing the lungs and heart and enclosed by the rib cage; the midbody section of an insect to which the legs and wings are attached.

Toxin
A poisonous substance produced by a plant or animal, which is often used as a means of defense. A toxin-producing animal secretes toxins within its own body.

Trachea
A tube in vertebrates that conducts air between the throat and bronchi; one of the tubes that form a system through which air travels to the cells of the body in insects and other land-living vertebrates.

Transpiration
Evaporation of water from the leaves of a plant, which helps draw water up the stem.

Tuber
A swollen underground part of a plant root—for example, a potato—that acts as a storage organ for starch.

Urea
A waste product formed when proteins are broken down in the liver. Urea is excreted in the urine.

Urine
The watery fluid produced in the kidneys that carries ammonia, uric acid, urea, amino acids, and other waste products from the body through the urethra or cloaca after being stored in the bladder.

Uterus
Hollow, muscular organ in which an embryo develops after implantation in the endometrium, or lining.

Vacuole
A membrane-bound cavity within a cell that may be used to store or digest food, store waste, or regulate the water content of the cell.

Vascular plant
Any plant that has conducting tissues, such as phloem and xylem, along which fluids can pass.

Vascular tissue
Type of conducting tissue found in vascular plants. It is made up of the xylem, phloem, sclerenchyma, and parenchyma cells.

Vein
A vessel that carries blood from the body to the heart. Veins almost always carry oxygen-poor blood, the exception being the pulmonary veins through which oxygenated blood from the lungs is pumped to the heart.

Vertebrate
An animal with a backbone. The five classes of vertebrates are amphibians, birds, fish, mammals, and reptiles.

Vestigial organ
Organ that has become reduced in size and structure over time because it is no longer required.

Vivipary
A method of reproduction in which the developing embryo obtains its nourishment directly from the mother, usually through a placenta. Viviparous animals include most mammals, and some fish, amphibians, reptiles, and insects.

Xylem
Vascular tissue in plants that conducts water and dissolved salts from the roots to other parts of the organism.

Resources for further study

Bibliography

Aquatic invertebrates

Anderson, D. T. (ed.) 2001. *Invertebrate Zoology*. Oxford University Press: New York.

Arai, M. N. 1997. *A Functional Biology of the Scyphozoa*. Chapman and Hall: London.

Blaxland, Beth. 2003. *Crabs, Crayfishes, and Their Relatives*. Chelsea House: Philadelphia, PA.

Blum, Mark. 1998. *Bugs in 3-D*. Chronicle: San Francisco, CA.

Brusca, R. C., and G. J. Brusca. 2003. *Invertebrates*. Sinauer: Sunderland, MA.

Factor, J. R. 1995. *The Biology of the Lobster, Homarus americanus*. Academic: New York.

George, T. C. 2001. *Jellies: The Life of Jellyfish*. Millbrook: Brookfield, CT.

Gittleman, A. M., and O. M. Amin. 2001. *Guess What Came to Dinner? Parasites and Your Health*. Penguin: New York.

Kraynak, Joe, and Kim W. Trenault. 2003. *The Complete Idiot's Guide to the Oceans*. Penguin: New York.

Petersen, Christine. 2002. *Invertebrates*. Scholastic Library: Danbury, CT.

Ruppert, Edward E., and Robert D. Barnes. 1994. *Invertebrate Zoology*. Saunders College: Fort Worth, TX.

Shick, J. Malcolm. 1991. *A Functional Biology of Sea Anemones*. Chapman and Hall: New York.

Birds

Bishop, N. 1997. *The Secrets of Animal Flight*. Houghton-Mifflin: Boston, MA.

Brennan, P. 2002. *Penguins and Other Flightless Birds*. Animals of the World. World Book: New York.

Cowan, Richard. 2000. *History of Life*. Blackwell Science: Boston, MA.

Lynch, Wayne. 1997. *Penguins of the World*. Firefly: Toronto, Canada.

Parry-Jones, J., and F. Greenaway. 2000. *Eyewitness: Eagles and Birds of Prey*. Dorling Kindersley: London.

Proctor, N. S., and Patrick J. Lynch. 1993. *Manual of Ornithology*. Yale University Press: New Haven, CT.

Schafer, Kevin. 2000. *Penguin Planet: Their World, Our World*. Northword: Chanhassen, MN.

Stone, Lynn, M. 2003. *Bald Eagles*. Lerner: Minneapolis, MN.

Winkler, Hans, David Christie, and David Nurney. 1995. *Woodpeckers: A Guide to the Woodpeckers, Piculets, and Wrynecks of the World*. Pica: Robertsbridge, UK.

Fish

Ashley, L. M. 1988. *Laboratory Anatomy of the Shark*. McGraw-Hill: Columbus, OH.

Forey, P. L. 1998. *History of the Coelacanth Fishes*. Chapman and Hall: London.

Hamlett, W. C. (ed.) 1999. *Sharks, Skates, and Rays: The Biology of Elasmobranch Fishes*. Johns Hopkins University Press: Baltimore, MD.

Helfman, Gene, Douglas Facey, and Bruce Collette. 1997. *The Diversity of Fishes*. Blackwell Science: Boston, MA.

Jackson, J. (ed.) 2004. *Encyclopedia of the Aquatic World*. Marshall Cavendish: Tarrytown, NY.

Kuiter, R. H. 2000. *Seahorses, Pipefishes, and Their Relatives*. TMC: Horleywood, UK.

Mallory, K. 2001. *Swimming with Hammerhead Sharks*. Houghton-Mifflin: Boston, MA.

Mayden, R. L. 1992. *Systematics, Historical Ecology, and North American Freshwater Fishes*. Stanford University Press: Stanford, CA.

Moyle, P. B. 1995. *Fish: An Enthusiast's Guide*. University of California Press, Berkeley, CA.

Moyle, P. B., and J. J. Cech. 2000. *Fishes: Introduction to Ichthyology*. Prentice Hall: Upper Saddle River, NJ.

Nelson, J. S. 1994. *Fishes of the World*. (3rd ed.) Wiley: New York.

Paxton, J. R., and W. N. Eschmeyer (eds.). 1998. *Encyclopedia of Fishes*. Academic: San Diego, CA.

Petrinos, C. 2001. *Realm of the Pygmy Seahorse: An Underwater Photography Adventure*. Starfish: Athens, Greece.

Randall, D. J., and A. P. Farrell (eds.). 1997. *Deep-Sea Fishes*. Academic: San Diego, CA.

Weinberg, S. 2000. *A Fish Caught in Time*. HarperCollins: London.

Mammals

Alderton, David. 1996. *Rodents of the World*. Facts On File: New York.

Ankel-Simons, Friderun. 1999. *Primate Anatomy: An Introduction*. Elsevier Academic: Boston, MA.

Baggaley, A., and J. Hamilton. 2001. *Human Body: An Illustrated Guide to Every Part of the Human Body and How It Works*. DK: New York.

Bonner, Nigel W. 1989. *The Natural History of Seals*. Christopher Helm: London.

Chivers, D. J., and P. Langer (eds.). 1994. *The Digestive System in Mammals: Food, Form, and Function*. Cambridge University Press: New York.

Conniff, Richard. 2002. *Rats! The Good, the Bad, and the Ugly*. Crown: New York.

Eckert, R. 1997. *Animal Physiology*. Freeman: New York.

Eltringham, S. K. 1999. *The Hippos*. Poyser: London.

Etses, R. D. 1991. *The Behavior Guide to African Mammals*. University of California Press: Berkeley, CA.

Evans, P. J. H. 2001. *Marine Mammals: Biology and Conservation*. Plenum: New York.

Geist, V. 1998. *Deer of the World: Their Evolution, Behavior, and Ecology*. Stackpole: Mechanicsburg, PA.

Grant, T. R. 1995. *The Platypus: A Unique Mammal*. University of New South Wales Press: Sydney, Australia.

Goodall, Jane. 1996. *My Life with the Chimpanzees*. Aladdin Paperbacks: New York.

Hall, Leslie. 2000. *Flying Foxes: Fruit and Blossom Bats of Australia*. Reed: Sydney, Australia.

Hare, T., and M. Lambert. 1997. *The Encyclopedia of Mammals*. Marshall Cavendish: New York.

Kardong, Kenneth V. 1995. *Vertebrates*. William C. Brown: Dubuque, IA.

Kitchener, A. 1991. *The Natural History of the Wild Cats*. Natural History of Mammals Series. Cornell University Press: Ithaca, NY.

Kunz, Thomas H., and M. Brock Fenton. 2003. *Bat Ecology.* University of Chicago Press: Chicago, IL.

Lazaroff, M. 2004. *The Complete Idiot's Guide to Anatomy and Physiology.* Penguin: New York.

Macdonald, David W. 2006. *The Encyclopedia of Mammals.* Facts On File: New York.

Macdonald, David W., and C. Sillero-Zubiri (eds.). 2004. *The Biology and Conservation of Wild Canids.* Oxford University Press: Oxford, UK.

McGowan, Christopher. 1999. *A Practical Guide to Vertebrate Mechanics.* Cambridge University Press: Cambridge, UK.

Mead, James G., and Joy P. Gold. 2002. *Whales and Dolphins in Question: The Smithsonian Answer Book.* Smithsonian Books: Washington, DC.

Neuweiler, Gerhard. 2000. *The Biology of Bats.* Oxford University Press: New York.

Nowak, Ronald M. 1999. *Walker's Mammals of the World.* Johns Hopkins University Press: Baltimore, MD.

Nowell, K., and P. Jackson (eds.). 1996. *Wild Cats.* IUCN: Gland, Switzerland.

Paxinos, George (ed.). 2004. *The Rat Nervous System.* Elsevier Academic: Boston, MA.

Perrin, W. F., B. Würsig, and J. G. M. Thewissen (eds.). 2002. *Encyclopedia of Marine Mammals.* Academic: San Diego, CA.

Purves, W. K., G. H. Orians, D. Sadava, and H. C. Heller. 2003. *Life: The Science of Biology.* Freeman: New York.

Raven, Peter H., George B. Johnson, Susan R. Singer, and Jonathan B. Losos. 2004. *Biology.* McGraw-Hill Science: New York.

Reynolds, John E., III, and S. A. Rommel (eds.). 1999. *Biology of Marine Mammals.* Smithsonian Institution Press: Washington, DC.

Rismiller, P. 1999. *The Echidna: Australia's Enigma.* Hugh Lauter Levin: Westport, CT.

Rue, L. L. 2004. *The Encyclopedia of Deer.* Voyageur: Stillwater, MN.

Russell, C., and M. Enns. 2003. *Grizzly Seasons: Life with the Brown Bears of Kamchatka.* Firefly: Richmond Hill, Ontario, Canada.

Schneider, B. 2003. *Where the Grizzly Walks: The Future of the Great Bear.* Falcon: Guilford, CT.

Silverthorn, Dee. 1998. *Human Physiology: An Integrated Approach.* Prentice Hall: Upper Saddle River, NJ.

Sunquist, M., and F. Sunquist. 2002. *Wild Cats of the World.* University of Chicago Press: Chicago, IL.

Swindler, Danis Ray. 2002. *Primate Dentition: An Introduction to the Teeth of Non-Human Primates.* Cambridge University Press: Cambridge, UK.

Tortora, G. J., S. R. Grabowski, and B. Roesch. 2000. *Principles of Anatomy and Physiology.* (9th ed.) John Wiley: New York.

Van der Graaf, K. 1997. *Schaum's Outline of Human Anatomy and Physiology.* McGraw-Hill: Columbus, OH.

Vaughan, Terry A. 1999. *Mammalogy.* Brooks/Cole: Belmont, CA.

Vogel, Steven. 2003. *Comparative Biomechanics: Life's Physical World.* Princeton University Press: Princeton, NJ.

Vrba, Elisabeth S., and George B. Schaller (eds.). 2000. *Antelopes, Deer, and Relatives: Fossil Record, Behavioral Ecology, Systematics, and Conservation.* Yale University Press: New Haven, CT.

Microorganisms

Flint, S. J., V. R. Racaniello, L. W. Enquist, and A. M. Skalka. 2003. *Principles of Virology: Molecular Biology, Pathogenesis, and Control of Animal Viruses.* American Society of Microbiology: Washington, DC.

Madigan, Michael T., J. M. Mantinko, and Jack Parker. 2002. *Brock's Biology of Microorganisms.* Prentice Hall: Upper Saddle River, NJ.

Singleton, Paul. 2004. *Bacteria in Biology, Biotechnology, and Medicine.* Wiley: New York.

Sompayrac, L. 2002. *How Pathogenic Viruses Work.* Jones and Bartlett: Boston, MA.

Srivastava, Sheela. 2003. *Understanding Bacteria.* Kluwer Academic: Boston, MA.

Plants, fungi, and seaweed

Anderson, E. F. 2001. *The Cactus Family.* Timber: Portland, OR.

Baskin, C. C., and J. M. Baskin. 1998. *Seeds: Ecology, Biogeography, and Evolution of Dormancy and Germination.* Academic: San Diego, CA.

Bell, Adrian D. 1991. *Plant Form: An Illustrated Guide to Flowering Plant Morphology.* Oxford University Press: Oxford, UK.

Bell, P. R., and A. R. Hemsley. 2005. *Green Plants: Their Origin and Diversity.* Cambridge University Press: New York.

Cafferty, S. 2005. *Firefly Encyclopedia of Trees.* Firefly: Toronto, Canada.

D'Amato, P. 1998. *The Savage Garden: Cultivating Carnivorous Plants.* Ten Speed: Berkeley, CA.

Dickinson, William C. 2000. *Integrative Plant Anatomy.* Elsevier Academic: Boston, MA.

Garnier, E., *et al.* (eds.) 1999. *Variations in Leaf Structure: An Ecophysiological Perspective.* Cambridge University Press: Cambridge, UK, and New York.

Graham. L. E., and L. Warren Cox. 2000. *Algae.* Prentice Hall: Upper Saddle River, NJ.

Hall, D. O., and K. K. Rao. 1999. *Photosynthesis.* (6th ed.) Cambridge University Press: New York.

Heiser, Charles B., Jr. 1987. *The Fascinating World of the Nightshades: Tobacco, Mandrake, Potato, Tomato, Pepper, and Eggplant.* Dover: New York.

Heywood, V. H. 2006. *Flowering Plants of the World.* Firefly: Toronto, Canada.

Hughes, M. S. 1998. *Buried Treasure: Roots and Tubers.* Lerner: Minneapolis, MN.

Jackson, J. E. 2003. *Biology of Apples and Pears.* Cambridge University Press: Cambridge, UK.

Jennings, D. H. 1999. *Fungal Biology: Understanding the Fungal Lifestyle.* Springer: New York.

King, J. 1998. *Reaching for the Sun: How Plants Work.* Cambridge University Press: Cambridge, UK.

Malcolm, W., et al. 2000. *Mosses and Other Bryophytes: An Illustrated Glossary.* Timber: Portland, OR.

Motley, T. J., N. Zerega, and H. Cross (eds.). 2005. *Darwin's Harvest: New Approaches to the Origins, Evolution, and Conservation of Crops.* Columbia University Press: New York.

Noble, P. S. 1994. *Remarkable Agaves and Cacti.* Oxford University Press: Oxford, UK.

Norstog, K. J., and T. J. Nicholls. 1997. *The Biology of the Cycads.* Cornell University Press: Ithaca, NY.

Pace, G. 1998. *Mushrooms of the World.* Firefly: Toronto, Canada.

Pietropaolo, J., and P. Pietropaolo. 1996. *Carnivorous Plants of the World.* Timber: Portland, OR.

Preston, R. 1995. *The Hot Zone.* Anchor: New York.

Proctor, M., P. Yeo, and A. Lack. 1996. *The Natural History of Pollination.* Timber: Portland, OR.

Simpson, B. B., and M. Connor Ogorzaly. 1986. *Economic Botany: Plants in Our World.* McGraw-Hill: New York.

Thomas, D. N. 2002. *Seaweeds.* Smithsonian Institution Press: Washington, DC.

Tyree, Melvin T., and M. H. Zimmermann. 2003. *Xylem Structure and the Ascent of Sap.* Springer: New York.

Van den Hoek, C., D. Mann, and H. M. Jahns. 1996. *Algae: An Introduction to Phycology.* Cambridge University Press: New York.

Weier, T. Elliot, C. Ralph Stocking, M. G. Barbour, and T. L. Rost. 1982. *Botany.* Wiley: New York.

Reptiles and amphibians

Halliday, T., and K. Adler. 2002. *The New Encyclopedia of Reptiles and Amphibians.* Firefly: Toronto, Canada.

Harris, T. (ed.) 2003. *Reptiles and Amphibians.* Marshall Cavendish: Tarrytown, NY.

Mattison, C. 1989. *Lizards of the World.* Facts On File: New York.

Mattison, C. 2003. *Snakes of the World.* Facts On File: New York.

McDiarmid, R. W., and R. Altig. 1999. *Tadpoles: The Biology of Anuran Larvae.* University of Chicago Press: Chicago, IL.

Murphy, John C., and Robert W. Henderson, 1997. *Tales of Giant Snakes: A Historical Natural History of Anacondas and Pythons.* Krieger: Melbourne, FL.

Orenstein, R., G. Zug, and J. Mortimer. 2001. *Turtles, Tortoises, and Terrapins.* Firefly: Toronto, Canada.

Pough, F. H., R. M. Andrews, J. E. Cadle, M. L. Crump, A. H. Savitzsky, and K. D. Wells. 2003. *Herpetology.* (3rd ed.) Prentice Hall: Upper Saddle River, NJ.

Smith, H. M. 1995. *Handbook of Lizards of the United States and Canada.* Cornell University Press: Ithaca, NY.

Weishampel, David B., Peter Dodson, and Halszka Osmolska. 2004. *The Dinosauria.* Salamander: London, UK.

Terrestrial invertebrates

Anderson, D. T. (ed.) 2001. *Invertebrate Zoology.* Oxford University Press: New York.

Austin, Andrew, and Mark Dowton. 2000. *Hymenoptera: Evolution, Biodiversity and Biological Control.* CSIRO: Melbourne, Australia.

Beverley, Claire, and David Ponsonby. 2003. *The Anatomy of Insects and Spiders: Over 600 Excquisite Forms.* Chronicle: San Francisco, CA.

Blum, Mark. 1998. *Bugs in 3-D.* Chronicle: San Francisco, CA.

Brusca, R. C., and G. J. Brusca. 2003. *Invertebrates.* Sinauer: Sunderland, MA.

Chapman, R. F. 1999. *The Insects: Structure and Function.* Cambridge University Press: Cambridge, UK.

Foelix, Rainer F. 1996. *Biology of Spiders.* Oxford University Press: New York.

Green, Rick. 2002. *Apis Mellifera: A.K.A. Honeybee.* Branden: Boston, MA.

Himmelman, John. 2002. *Discovering Moths.* Down East: Camden, ME.

Hölldobler, Bert, and Edward O. Wilson. 1995. *Journey to the Ants*. Barnes and Noble: New York.

Jackson, T., and J. Martin (eds.). *Insects and Spiders of the World*. 2003. Marshall Cavendish: Tarrytown, NY.

Kitching, Ian J., and Jean-Marie Cadiou. 2000. *Hawkmoths of the World*. Cornell University Press: Ithaca, NY.

McGavin, G. C. 2000. *Insects, Spiders, and Other Terrestrial Arthropods*. Dorling Kindersley: New York.

Petersen, Christine. 2002. *Invertebrates*. Scholastic Library: Danbury, CT.

Ruppert, Edward E., and Robert D. Barnes. 1994. *Invertebrate Zoology*. Saunders College: Fort Worth, TX.

Scott, James, A. 1992. *The Butterflies of North America*. Stanford University Press: Stanford, CA.

Zahradník, J. 2000. *Bees and Wasps*. Silverdale: Leicester, UK.

Systems, cell biology, and genetics

Alberts, B., A. Johnson, J. Lewis, M. Raff, K. Roberts, and P. Walter. 2002. *Molecular Biology of the Cell*. Garland: New York.

Arnold, Nick. 1999. *Horrible Science: Disgusting Digestion*. Scholastic Library: Danbury, CT.

Ballard, C. 2002. *The Lungs and Breathing*. Heinemann Library: Crystal Lake, IL.

Bastian, G. 1997. *An Illustrated Review of the Skeletal and Muscular Systems*. Addison–Wesley: Boston, MA.

Beckingham, I. J. 2001. *ABC of Liver, Pancreas, and Gallbladder*. BMJ: Philadelphia, PA.

Callaghan, C. A., and B. M. Brenner. 2000. *The Kidney at a Glance*. Blackwell Science: Boston, MA.

Friedlander, Mark, and Terry M. Phillips. 1998. *The Immune System: Your Body's Disease-Fighting Army*. Lerner: Minneapolis, MN.

Futuyma, D. 1998. *Evolutionary Biology*. Sinauer: Sunderland, MA.

Gould, Stephen J. (ed.) 2001. *The Book of Life: An Illustrated History of the Evolution of Life on Earth*. Norton: New York.

Griffin, J. E., and S. R. Ojeda. 1996. *Textbook of Endocrine Physiology*. (3rd ed.) Oxford University Press: Oxford, UK.

Harold, Franklin. 2001. *The Way of the Cell: Molecules, Organisms, and the Order of Life*. Oxford University Press: Oxford, UK.

Hickman, B. F. 2001. *Perception: The Amazing Brain*. Blackbirch: New York.

Marshall Graves, Jenny. 2004. *Sex, Genes, and Chromosomes*. Cambridge University Press: Cambridge, UK.

Overy, Angela. 1997. *Sex in Your Garden*. Fulcrum: Golden, CO.

Restak, R. M. 2001. *The Secret Life of the Brain*. National Academy Press: Washington, DC.

Seibel, M. J., *et al.* (eds.) 1999. *Dynamics of Bone and Cartilage Metabolism*. Academic: New York.

Siegal, I. S. 1998. *All about Bone: An Owner's Manual*. Demos Medical: New York.

Snedden, R. 2003. *Cell Division and Genetics*. Heinemann Library: Chicago, IL.

Sompayrac, Lauren. 2003. *How the Immune System Works*. Blackwell: Malden, MA.

Teaford, M. F., M. M. Smith, and M. W. J. Ferguson. 2000. *Development, Function, and Evolution of Teeth*. Cambridge University Press, Cambridge, UK.

Internet resources

A Checklist of Amphibian Species and Identification Guide
Covers all the North American amphibian species.
www.npwr.usgs.gov/narcam/idguide

Albatross Project
Albatross ecology and conservation.
www.wfu.edu/albatross/index.htm

American Museum of Natural History
Resource that includes a virtual tour of the museum.
http://www.amnh.org

Amphibiaweb
Database covering all amphibians.
http://amphibiaweb.org

Animal Diversity Web
Information about the taxonomy and characteristics of animals.
http://animaldiversity.ummz.umich.edu/site/index.html

Arizona–Sonora Desert Museum
Includes information on cacti.
www.desertmuseum.org/books/cacti.html

Avibase
Database on almost 10,000 species of birds.
http://www.bsc-eoc.org/avibase/avibase.jsp

Carnivorous Plants:
A website devoted to pitcher plants and other carnivorous plant species.
www.waynesword.palomar.edu/carnivor.htm

CAS Echinoderm Webpage:
Site with links to many echinoderm-related resources.
www.calacademy.org/research/izg/echinoderm

Cells Alive
Resource on cell biology microbiology, immunolony, and microscopy.
http://www.cellsalive.com

Cephbase
Database covering all cephalopods.
http://www.cephbase.utmb.edu

Comparative Mammalian Brain Collections
Pictures of brains and brain slices from more than 100 species of mammals.
http://www.brainmuseum.org

EMBL Reptile Database
Large resource devoted to reptiles.
http://www.emblheidelberg.de/~uetz/LivingReptiles.html

ENature
Database of more than 5,500 species of animals and plants.
http://www.enature.com

Evolution
Information about all aspects of evolution.
http://www.pbs.org/wgbh/evolution

Field Guide to Boll Weevil Identification
Illustrated guide to common crop pests.
www.msucares.com/pubs/techbulletins/tb0228.pdf

Fishbase
Database of all known fish.
http://www.fishbase.org

General Crustacea Information:
An information retrieval system for the world's crustaceans.
www.crustacea.net

Hall of Mammals
Information and links to many mammal sites.
dwww.ucmp.berkeley.edu/mammal/mammal.html

How Animals Work
Animation showing how birds' lungs function.
www.sci.sdsu.edu/multimedia/birdlungs

How Your Immune System Works
Information on the human immune system.
http://health.howstuffworks.com/immune-system.htm

Human Anatomy Online
Resource showing anatomy of human body systems.
http ://www.innerbody.com/htm/body.html

Immune system
Information on the immune system and infectious diseases.
www.niaid.nih.gov/final/immun/immun.htm

International Carnivorous Plant Society
Links to sites about carnivorous plants.
www.carnivorousplants.org

Iowa State Entomology Index of Internet Resources
A database of insect-related sites.
www.ent.iastate.edu/list

JGI Center for Primate Studies
Information on the immune system of chimpanzees.
www.discoverchimpanzees.org

Lobster Conservancy
Resource on lobsters.
www.lobsters.org

Museum of Vertebrate Zoology, University of California
Resource showing collections of the museum.
http://mvz.berkeley.edu

National Geographic
The Web site of the National Geographic Society.
http://www.nationalgeographic.com

Natural History Museums
Web links to natural history museums and collections around the world.
http://www.lib.washington.edu/sla/natmus.html

Natural Perspective
A collection of images of four of the kingdoms: protists, fungi, plants, and animals.
http://perspective.com/nature

NetFrog Page
Online dissection of a virtual frog.
http://curry.edschool.virginia.edu/go/frog

Neuroscience for Kids
Information about the nervous system, with activities and experiments.
http://faculty.washington.edu/chudler/neurok.html

Operation Rubythroat
Resources on ruby-throated hummingbirds.
www.rubythroat.org

Plant Anatomy and Glossary
Information on plant structure.
http://dallas.tamu.edu/weeds/anat.html

Rocky Intertidal Shores
Resource on the life-forms of an important habitat.
www2.mcdaniel.edu/Biology/wildamerica/rockyforweb/rockyshores.html

ScienceDaily
Resource giving the latest scientific news stories.
http://www.sciencedaily.com

Seahorse.org
Information relating to sea horse anatomy.
www.seahorse.org

Smithsonian National Museum of Natural History
Online information about the natural history museum.
http://www.mnh.si.edu

Squirrel Page
A squirrel information resource.
www.members.tripod.com/Thryomanes/squirrel

Tree of Life
Over 1,350 Web pages on the diversity of life.
http://tolweb.org/tree/phylogeny.html

***Vaejovis carolinianus* (Carolina Scorpion)**
www.lander.edu/rsfox/310vaejovisLab.htm

Vertebrate Zoology
Web links to information about vertebrate animals.
http://www.lions.odu.edu/~kkilburn/vzhome.htm

Virtual Canine Anatomy
Information about dog anatomy.
www.cvmbs.colostate.edu/vetneuro

What Is Photosynthesis?
Resource on different aspects of photosynthesis.
http://photoscience.la.asu.edu/photosyn/education/learn.html

World Wildlife Fund
Information about endangered wildlife.
http://www.worldwildlife.org

Index of behaviors, habitats, and ecology

Volume numbers are in **bold** type. Page numbers to main articles are in **bold**; those in *italics* refer to captions or to names that occur in family trees.

A

altruism, in wolves **10**:1401
anadromous species, lampreys **4**:448
autotomy **2**:237; **5**:629

B

bathypelagic zone, gulper eels **3**:439
biological control, of rabbits **4**:488
bioluminescence **3**:439, *440, 444*; **10**:1348–1351, 1354
burrowing
  bivalves **3**:*347*, 351
  hagfish **4**:457
  snakes, skulls of **3**:*403*

C

calls, frog **1**:*55*, 57, 61
camouflage
  caterpillars **4**:499
  chameleons **5**:630
  coelacanths **1**:*123*
  deer **7**:969
  giraffes **3**:359
  hares **4**:*481*
  lions **5**:669
  lobsters **5**:686
  octopuses **6**:815–816
  porcupines **7**:902
  sea horses **8**:1100
  sloths **9**:1183
  snakes **3**:400
  tigers **5**:668

trout **9**:1282–1283
zebras **10**:1421
see also countershading
coevolution *see* evolution, convergent
color change
  chameleons **5**:*631*; **6**:*789*
  octopuses, squid, and cuttlefish **6**:815–*816*
  sailfish **8**:1064
communication
  crocodile **2**:152
  elephant **2**:*253*, 257, 262, 263
  giraffe **3**:366
  honeybee **4**:534
  hyena **5**:600–*601*
  kangaroo **5**:657
  lions **5**:669, 676
  manatee **5**:710
  whale **2**:201
  wolf **10**:1388, *1389*, 1401–1402
  zebra **10**:*1420*
  see also calls; scent
coprophagy **2**:*182*; **4**:*491*; **7**:961
coral reefs **8**:1087; **9**:1167
  giant clams **3**:*343, 344*
countershading
  sailfish **8**:1063
  whales **3**:379
courtship
  crab **1**:135, 140, *141*
  crocodile **2**:162–163
  hippopotamus **4**:519
  lobster **5**:691
  newt **6**:*810*–811
  octopus **6**:823
  porcupine **7**:914
  rhino **8**:1047
  scorpion **8**:*1082*

sea horse **8**:1109
spider **9**:1264, 1265
stingray **9**:1251
turtles **9**:1206
cud, chewing **10**:1375–1376

D

deep-sea environment *see* ocean depths
defense
  giraffe **3**:360
  millipede **3**:*340*
  sea anemone **8**:1090
  tarantula **9**:1256
  tortoise **9**:*1268*
deserts, antelope jackrabbits **4**:479
diving
  manatees **5**:712
  platypuses **7**:893
  seals **8**:1117, 1121
  swordfish **8**:1064
  whales **2**:202, 203

E

echolocation
  bats **3**:305
  whales and dolphins **2**:189, 193, 201
epiphytes
  cacti as **8**:1050
  orchids **6**:825, 826, 827, 830, *831*
estivation, amphibians **8**:1030
evolution, convergent
  ant-eating mammals **3**:315
  arthropod eyes **3**:337
  beak of echidnas and birds **7**:889
  between octopuses and vertebrates **6**:820

bill of platypuses and birds **7**:889
  deep-sea fish **3**:440
  euphorbias and cacti **8**:1052, *1054*
  hawkmoth ears **4**:498
  jerboas and kangaroo rats **7**:951
  Old World and New World porcupines **7**:901
  orchids and insects **6**:832
  and swimming underwater **3**:379
"evolutionary arms race" **5**:608

F

'fight or flight' response **4**:570; **10**:1372–1373
fighting for mates
  deer **7**:982
  giant tortoises **9**:*1271*
  giraffes **3**:364
  hares **4**:493
  sea horses **8**:1108
  weasels **10**:1341
fighting for territory, wildebeests **10**:*1379*
filter feeders **2**:175
flehmen response **4**:519; **5**:676; **7**:976; **10**:1373, *1430*
forests, pygmy hippopotamuses **4**:510

G

geophagia **10**:1434

H

herbivores
  intestines **2**:181
  teeth **2**:174

hibernation
  bears
    grizzly **3**:425, 427, 428,
      429
    polar **3**:*425*
  tortoises **9**:1276

I

infanticide, lions **5**:682
inflation, chameleons
  5:632
infrasound
  giraffe **3**:366
  hippopotamus **4**:519
  manatee **5**:710
ink, of squid, cuttlefish,
  and octopuses
  6:*817*
islands, animals on **4**:511

L

light production *see*
  bioluminescence

M

mate-guarding, grizzly
  bears **3**:428
migration
  billfish **8**:1073
  crabs **1**:140
  gray whales **3**:395
  hummingbirds **5**:590
  trout **9**:1291
  wildebeest **10**:*1364*
mimicry
  of bees **4**:530
  by caterpillars **4**:499
mutualism, between
  giant clams and
    zooxanthellae **3**:352
myxomatosis **4**:*488*

N

navigation, baleen whales
  3:385

O

ocean depths
  fish **3**:439, 440
  gulper eels **3**:*437*, 439
omnivores, teeth **2**:174

P

parasites
  on bovids **10**:1374
  lampreys **4**:447
  lice **5**:692, 693
  ticks **5**:698
  viroids **10**:1317
parental care
  centipedes **3**:341
  frogs and toads **1**:63
  octopuses **6**:823
  scorpions **8**:*1083*
  sea horses **8**:1109
phoresy **8**:1075
play
  dolphin **2**:*207*
  wolf **10**:*1403*
polygamy, squirrels
  9:1222
porpoising **2**:*192*;
  8:*1031*
pronking **10**:1371

R

regurgitation
  by eagles **2**:230
  by wolves **10**:1399,
    1402, 1403
  by woodpeckers
    10:1415, *1417*
rivers and lakes, common
  hippopotamuses
  4:510–511

S

savanna, giraffes **3**:366
scent
  cat **7**:939
  deer **7**:969, *976*
  giant anteater **3**:327

hyena **5**:595, 600–*601*
manatee **5**:716
moose **7**:*967*
otter **6**:854, 860
porcupine **7**:909
rat **7**:956, 963
weasel **10**:1335
wildebeest **10**:*1363*
wolf **10**:1399
seabed
  angel sharks **4**:463
  hagfish **4**:547
singing, whales **3**:386,
  394
sleep, ruminants and
  10:1373
sound production
  baleen whales **3**:388
  see also infrasound
spyhopping, whales
  3:386
swimming
  dolphins **2**:192, 197,
    198
  fish **9**:1285
    billfish **8**:1066–1067
    coelacanths **1**:*125,
      127*
    guitarfish **9**:1243
    sharks **4**:464
    skates and rays
      9:*1243–1244*
    tuna **6**:764–765;
      8:1066–*1067*
  manatees **5**:708
  penguins **7**:*870–871,*
    873–874
  underwater **3**:379;
    6:848
  whales **2**:198; **3**:376,
    379, 384
symbiosis
  ants "farming" other
    insects **1**:33
  and the evolution of
    eukaryotes **2**:170

light-producing
  bacteria in fish
  3:439
microorganisms in guts
  of lice **5**:697–698
  of sloths **9**:1191
zooxanthellae in
  cnidarians **8**:1092

T

thigmotropism **9**:1161
tool-use
  chimpanzees and **1**:97,
    104
  elephants and **2**:257
trophobiosis **1**:33

U

Unkenreflex **6**:804

V

viral hemorrhagic disease
  (VHD) **4**:488

W

warning coloration
  honeybees **4**:530
  newts **6**:804
  snakes **3**:400
  weasels **10**:1330–1331

Index of biological classification

Amphibians and reptiles

B

bullfrog 1:*54–63*
  calls 1:*57*, 61, 63
  circulatory and respiratory systems 1:55, *61*
  digestive and excretory systems 1:55, *62*
  external anatomy 1:55, *56–57*
  nervous system 1:55, *60*
  reproductive system 1:55, *63*
  skeletal and muscular systems 1:55, *58–59*

C

caecilians 1:*54*; 3:407; 6:*800*
caiman 2:*148*, 149
  dwarf 2:*148*, 149
  nostrils 2:*160*
  snout 2:150
  spectacled 2:*160*
chameleon 5:*624*, 625
  circulatory and respiratory systems 5:625, *632*
  color change 5:*631*; 6:*789*
  digestive and excretory systems 5:625, *633*
  dwarf 5:*624*
  external anatomy 5:625, *626–627*
  Hoehnel's 5:*634–635*
  Jackson's 5:*624–635*; 7:*997*
  muscular system 5:625, *630*
  nervous system 5:625, *631*

reproductive system 5:625, *634–635*; 7:*997*
  skeletal system 5:625, *628–629*
chelonians 9:1267
cobra, fangs 3:*404*
crocodile 2:*148–163*
  African dwarf 2:149
  American 2:*148*, 149, *150*
    blood and air flow 2:*159*
    brain and spinal cord 2:*158*
    digestive system 2:*161*
    jaws 2:153
    muscles 2:*156*
    palatal and nasal valves 2:*160*
    skeleton 2:*154*
  ancient 2:155
  biggest ever 2:154
  biting and swallowing 2:156–157, 161
  brain 2:158
  breathing 8:1028
  breathing underwater 2:160
  circulatory system 1:*112*; 2:149, 159–160
  communicating by smell 2:152
  differs from alligators 2:151
  digestive and excretory systems 2:149, 161
  dwarf 2:148
  ears 2:151
  eggs 2:162, 163
  external anatomy 2:149, 150–152
  eyes 2:150–151, 158
  jaws 2:151, 153, 158
  Johnston's 2:156

  moving and swimming 2:150, 156, 157
  muscular system 2:149, 156–157
  nervous system 2:149, 158
  Nile 2:*157*
  nostrils 2:150, 151
  reproductive system 2:149, 162–163
  respiratory system 2:149, 160
  ripple detectors 2:158
  salt glands 2:*152*, 161
  skeletal system 2:149, 153–155
  skull 2:153
  tail 2:157
  teeth 2:151, 153
  true 2:148, 149
crocodilians 2:149; 5:*624*; 9:*1194*
  scales 2:151–152
  snouts 2:150

D

dinosaurs, skeleton 9:1160

F

file snake 3:*401*
  Arafura 3:*399*
frog 1:*54*; 6:*800*
  "ancient" 1:*54*
  calls 1:*55*, 57, 61
  common 1:*54*
  Darwin's 1:*54*, 63
  edible 1:*54*
  European tree 1:*62*
  gold 1:*54*
  goliath 1:55
  jumping and swimming 1:*59*
  metamorphosis 1:57; 2:267
  "new" 1:*54*
  poison arrow 2:*265*

reproduction, external fertilization 7:*992*
  Rio Grande leopard 1:*54*
  skeleton 9:1159–1160
  tropical 1:*54*
  true 1:*54*, 55

G

gecko, skin color 5:*631*; 6:*789*
gharial 2:149
  false 2:150
  snout 2:150, *151*, 153
  teeth 2:*154*
green anaconda **3:*396–409***
  circulatory and respiratory systems 3:397, *407*
  digestive and excretory systems 3:397, *408*
  external anatomy 3:397, *398–401*
  muscular system 3:397, *405*
  nervous system 3:397, *406*
  reproductive system 3:397, *409*
  skeletal system 3:397, *402–404*; 9:*1175*

H

hellbender 6:801, 809

I

ichthyosaur 3:379
iguana
  marine, salt-secreting glands 2:161
  scales 5:*627*
  teeth 2:*174*

J

Jackson's chameleon **5:*624–635***

K

krait, venom **6**:783

L

lizard **3**:*396*; **5**:*624*, 625;
   9:*1194*
  alligator, scales **2**:151
  color change **6**:*789*
  limbless **3**:407
  red blood cells **4**:*521*
  sungazer, pheromones
    2:274
  tail that can drop off
    5:629
  vision **9**:1274
  whiptail, asexual
    reproduction **7**:*986*,
    989

M

matamata **9**:1198
mud puppy **6**:801, 803

N

newt **6:800–811**
  circulatory and
    respiratory systems
    6:801, *808–809*
  crested **6**:*802–803*,
    804, *805*, 807, *809*,
    810, 811
  external anatomy
    6:801, *802–804*
  larvae **6**:804
  nervous system **6**:801,
    807
  red-spotted **6**:*810*
  regrowing damaged
    parts **6**:803
  reproductive system
    6:801, *810–811*
  ribbed **6**:*806*
  skeletal and muscular
    systems **6**:801,
    805–806
  walking on land **6**:805

O

olm **6**:801

P

padloper, speckled
    9:1279
pterosaur, wings **3**:*299*
python **3**:*396, 400*
  heat-detection **3**:*406*
  keeping eggs warm
    3:405

R

ranids **1**:55
rattlesnake **3**:401
  heat-detection **3**:406
reptiles **2**:148–149;
    3:*396*; **5**:*624*;
    9:1195, *1266*, 1267
  brain **6**:*795*
  calcium regulation
    2:268
  excretion by **2**:186
  heart **1**:*110*, 112
  internal fertilization
    9:1279
  respiratory systems,
    ventilation
    8:1027–1028
  in salt water **2**:161
  skeletal system **9**:1160
  jaw **9**:1170

S

salamander **1**:*54*; **6**:*800*
  alpine **6**:802
  Asian land **6**:801
  Chinese giant **6**:801
  circulatory and
    respiratory systems
    6:*808–809*
  European **6**:*800*, 801
  fire **6**:802, 804, 807
  Japanese giant **6**:801
  larvae **6**:*804*
  lungless **6**:*800*, 801,
    806, 808

mole **6**:*800*, 801, *803*,
    804
movement **6**:805, 806
neoteny **6**:811
nervous system **6**:*781*
northern red **6**:*801*
Pacific giant **6**:801
senses **6**:807
skeleton **6**:805, 806;
    9:1159
spotted **6**:*803*
tiger **6**:801, *806*, 811
torrent **6**:801
siren **6**:800, 803
snake **3**:*396*, 397; **5**:*624*;
    9:*1194*
  blind **3**:*403*
  file *see* file snake
  pipe **3**:*403*
  sex determination
    9:1207
  teeth **2**:174
  thread **3**:*403*
  venom **2**:275
  vision **9**:1274
snapping turtle
    9:1194–1207, *1266*
  alligator
    9:*1195,1197–1200*,
    1205, 1206, 1207
  circulatory and
    respiratory systems
    9:1195, *1203–1204*
  common
    9:*1196–1197*, 1206,
    1207
  digestive and excretory
    systems **9**:1195,
    1205
  external anatomy
    9:1195, *1196–1198*
  muscular system
    9:1195, *1201*
  nervous system **9**:1195,
    1202
  reproductive system
    9:1195, *1206–1207*

skeletal system **9**:1195,
    1199–1200
squamates **3**:*396*; **5**:*624*;
    9:*1266*

T

tadpoles **1**:57, 61, 62,
    63
  metamorphosis **1**:57,
    62, 63; **2**:267
  newt **6**:802–803
terrapin
  diamond-backed
    9:1207
  painted **9**:1204
toad **1**:*54*, 55; **6**:*800*
  cane, poison glands
    2:*275*
  European common
    1:62
  Suriname **1**:63
  true **1**:*54*
tortoise **9**:*1194*, 1195,
    1266–1279
  Aldabra giant **9**:*1267*,
    1277, 1279
  circulatory and
    respiratory systems
    9:1267, *1275–1276*
  desert **9**:*1271*
  digestive and excretory
    systems **9**:1267,
    1277
  external anatomy
    9:1267, *1268–1271*
  Galápagos giant
    9:*1269*, 1271–1279
  shell shapes **9**:*1270*
  hinge-back **9**:1268,
    1273
  Horsfield's/Russian
    9:*1268*
  land **9**:1267, 1273,
    1277
  Madagascar plowshare
    9:1279
  Mediterranean **9**:1273

tortoise *(continued)*
  muscular system
      9:1267, *1273*
  nervous system **9**:1267,
      1274
  pancake **9**:1272,
      1278–1279
  reproductive system
      9:1267, *1278–1279*
  skeletal system **9**:1267,
      1272
  spurred **9**:*1279*
tuatara, pineal gland
      (parietal eye; third
      eye) **2**:273
turtle **5**:*624*; **9**:*1194,*
      1195, *1266*, 1267
  big-headed **9**:*1194,*
      1266
  box **9**:1198, 1204, 1268
  breathing **8**:1028
  Chinese softshell
      9:*1201*
  eastern box **9**:1207
  green sea **9**:*1276*
  hidden-necked **9**:*1194,*
      1195, *1266*, 1273
  leatherback, throat
      2:*176*
  mud **9**:1198
  musk **9**:1198
  plastrons **9**:*1198*
  pond **9**:*1198*, 1277
  sea **9**:*1194, 1266*
  senses **9**:1274
  shell **2**:152; **9**:*1270*
  side-necked **9**:*1194,*
      1266, 1273
  snapping *see* snapping
      turtle
  softshell **9**:*1194, 1266,*
      1267, 1275
  spiny softshell **9**:*1200*
  teeth **2**:174
  withstanding cold
      temperatures
      9:*1204*

V
viper **3**:401
  heat-detection **3**:406

W
worm-lizards
      (amphisbaenians)
      3:*407*; **5**:*624*

Birds

A
albatross **1**:**10–19**
  bill **1**:*11, 13*, 18
  circulatory and
      respiratory systems
      1:11, *17*
  digestive and excretory
      systems **1**:11, *18*
  external anatomy **1**:*11,*
      12–13
  flight **1**:*12*–13, *15*
  gray-headed **1**:*10*
  Hawaiian black-footed
      1:*16*
  Laysan **1**:*10*, 16, *19*
  long journeys **1**:13
  muscular system **1**:11,
      15
  reproductive system
      1:11, *19*
  royal **1**:*10*
  senses **1**:16
  short-tailed **1**:*10*
  skeletal system **1**:11, *14*
  sooty **1**:*10*, 11
  taxonomy **1**:*10*
  vision and eyesight
      1:*13*, 16
  wandering **1**:*10, 12,*
      13, 14, 15
  yellow-nosed **1**:*10*
auk, wing bones **7**:*874*

B
barbet **10**:*1404*
bataleur **2**:224

beaks **9**:1170
bones **1**:14; **9**:1168
brain **2**:229; **6**:*795*
buzzard
  common, eyes **2**:229
  flight **2**:223

C
calcium regulation **2**:268
cassowary **6**:*834*, 835

D
digestion by birds **2**:178
diving petrel **1**:*10*, 11
duck **5**:695

E
eagle **2**:**220–233**
  ancestors **2**:223
  bald **2**:*220*, 221, *222*
      brain and spine **2**:*229*
      guts and kidneys
          2:*230*
      muscles **2**:*227*
      respiratory system
          8:*1019*
      skeleton **2**:*225*
      skull **9**:*1170*
      toes **2**:224
      wings **3**:*299*
  bill **2**:*223*–224, 226,
      228, 230
  booted (true eagles)
      2:*220*, 221
  circulatory and
      respiratory systems
      2:221, 231
  digestive and excretory
      systems **2**:221, 230
  eggs **2**:232–233
  external anatomy
      2:221, *222–224*
  eyes **2**:224, 229
  feathers **2**:222, 223,
      228
  fish **2**:*220*, 221
      toes **2**:224

flight **2**:222, 227, 231
  Haast's **2**:223
  harpy **2**:*220*, 221;
      7:*902*
  legs and feet **2**:224, 226
  life expectancy **2**:*233*
  muscular system **2**:221,
      227–228
  neck **2**:224, 226, 228
  nervous system **2**:221,
      229
  regurgitation by **2**:230
  reproductive system
      2:221, 232–233
  skeletal system **2**:221,
      225–226
  snake **2**:*220*, 221, 224
  tail **2**:224, 225
  temperature regulation
      2:231
  wings **2**:222, 223
      bones **2**:*226*
      muscles **2**:227–228
elephant bird **6**:837
embryo **7**:*1005*
emu **6**:*834*
evolution from dinosaurs
      9:*1160*
excretion by **2**:186

F
falcon **2**:*220*
  African pygmy, feet
      2:*226*
flamingo **5**:695
flicker
  Andean **10**:1408
  gilded **8**:*1057*
  northern **10**:*1412*
flight **1**:13

G
gull, wing bones **7**:*874*

H
hawk **2**:*220*
hearing by birds **9**:1170

hummingbird
        5:*580–591*; **8**:1059
    Anna's **5**:*584*, 590
    black-chinned **5**:*583*
    circulatory and
        respiratory systems
        5:581, *588*
    digestive and excretory
        systems **5**:581,
        589–590
    external anatomy
        5:581, *582–584*
    flight **5**:*583–584*, 586
    giant **5**:582
    hermit **5**:*580*, 581
    muscular system **5**:581,
        585
    nervous system **5**:581,
        587
    reproductive system
        5:581, *591*
    ruby-throated **5**:*580*,
        581, *582*, 584, *585*,
        586, *588*, *590*, *591*
    skeletal system **5**:581,
        585
    sword-billed **5**:589

K

kite **2**:*220*
kiwi **6**:*834*, 835, 840

M

moa **2**:223; **6**:837

O

osprey **2**:*220*
ostrich **6**:*834–843*
    circulatory and
        respiratory systems
        6:835, *841*
    digestive and excretory
        systems **6**:835, *842*
    external anatomy
        6:835, *836–837*
    extinct relatives
        6:837

keeping cool **6**:841
    muscular system **6**:835,
        839
    nervous system **6**:835,
        840
    reproductive system
        6:*835, 843*
    skeletal system **6**:835,
        838
        hind limb bones
            9:*1175*
    two species of? **6**:837
oxpecker **8**:*1035*

P

penguin **3**:379;
        7:*868–881*
    Adélie **7**:*868*, 870, 873,
        874, 880
    blue **7**:*868*, 870
    chinstrap **7**:*868*, 873
    circulatory and
        respiratory systems
        7:869, *877–878*
    crested **7**:*868*
    digestive and excretory
        systems **7**:869, *879*
    diving by **7**:878
    emperor **7**:869, 870,
        871, 872, 873, 876,
        877, *879, 881*
    evolution **1**:14; **7**:*874*
    external anatomy
        7:869, *870–872*
    feathers **7**:871–872,
        875, 881
    Fiordland crested **7**:880
    Galápagos **7**:870, 877
    gentoo **7**:*868*, 873, 876
    Humboldt **7**:877
    king **7**:869, 870, 876,
        880, 881
    Magellanic **7**:881
    muscular system **7**:869,
        875
    nervous system **7**:869
    origins **7**:*872*

reproductive system
        7:869, *880–881*
    salt glands **7**:879
    skeletal system **7**:869,
        873–874
    swimming and
        porpoising
        7:*870–871*,
        873–874
    in the tropics **7**:877
    yellow-eyed **7**:*868, 869*
petrel **1**:*10*, 11
    see also diving petrel;
        storm petrel
piculet **10**:*1404*, 1405,
        1408, 1409, 1411
    olivaceous **10**:*1409*,
        1411
pigeon
    digestive system **2**:*178*
    magnetic-field
        detection **6**:791
    of prey **2**:220

R

ratites **6**:*834*
razorbill, wing bones
        7:*874*
respiratory systems in
        birds **8**:*1019*
    ventilation **8**:1028
rhea **6**:*834*, 835

S

sapsucker **10**:1415
    yellow-bellied **10**:1406,
        1415
secretary bird **2**:*220*
shearwater **1**:*10*
skeletal system in birds
        9:1160, 1175
    jaw **9**:1170
song **7**:*999*
sparrow, Lincoln's **7**:*999*
stork **5**:695
storm petrel **1**:*10*, 11
swift **5**:*580*

T

tinamou **6**:*834*, 835
toucan **10**:*1404*
tube-nosed seabirds **1**:*10*,
        11
    dimethyl sulfide and
        1:16
    evolution **1**:14
    salt glands **1**:18

V

vulture
    flight **2**:223, 231
    Rüppell's griffon
        2:231

W

woodpecker
        10:*1404–1417*
    acorn **10**:*1408*
    avoids brain damage
        10:*1413*
    black **10**:*1417*
    circulatory and
        respiratory systems
        10:1405, *1414*
    crimson-crested
        10:*1409*, *1415*
    digestive and excretory
        systems **10**:1405,
        1415
    external anatomy
        10:1405,
        1406–1409
    gila **8**:1057; **10**:*1405*
    great spotted **10**:*1407*,
        1409, 1410, 1412,
        1414
    green **10**:*1416*
    hairy **10**:*1406*
    Hispaniolan **10**:1416
    imperial **10**:*1406*
    making a noise
        10:*1414*
    muscular system
        10:1405, *1412*

woodpecker *(continued)*
  nervous system
      10:1405, *1413*
  pileated **10**:1408–1409,
      1410, 1417
  reproductive system
      10:1405,
      1416–1417
  rufous **10**:*1415*
  skeletal system
      10:1405,
      1410–1411
  three-toed **10**:1409
  true **10**:*1404,* 1405
wryneck **10**:*1404,* 1405,
      1409, 1415

Fish

A

angelfish, queen,
      changing sex
      7:998
angler fish, deep-sea
      3:440, *444*

B

barracuda **8**:*1060*
billfish **8**:1061, 1062,
      1064
  migration **8**:1073
  swimming
      8:1066–1067
bonefish **3**:*436*

C

carp, scales **9**:*1283*
catshark **4**:*460*
  barbeled **4**:*460*
  false **4**:*460*
  finback **4**:*460*
ceratodonts **1**:*122*
char **9**:1281
chimera **4**:*470*
clown fish, changing sex
      7:*998*

coelacanth **1:122–131**
  African **1**:*122, 124,*
      125, 127, 128, 129,
      130
  buoyancy **1**:129
  circulatory and
      respiratory systems
      1:123, *129*
  digestive and excretory
      systems **1**:123, *130*
  external anatomy
      1:123, *124–125*
  Indonesian **1**:*122,* 124
  muscular system **1**:123,
      127
  nervous system **1**:123,
      128
  reproductive system
      1:123, *131*
  scales **9**:1283
  skeletal system **1**:123,
      126
  swimming **1**:*125,* 127

D

dogfish, egg sacs **7**:*1004*

E

eel **4**:*436*
  boatail snipe (arrow)
      4:*436*
  deep-sea **4**:*436*
  gulper *see* gulper eel
  pelican **4**:*436,* 438, 444
      gills **4**:*443*
      head pores **4**:*442*
  single-jaw **4**:*436*
      larvae **4**:445
  swallower **4**:*436,* 437
elephant fish **4**:*470*

F

fish
  bony **1**:*122;* **3**:*436;*
      8:*1060,* 1061,
      1096, 1097;
      9:*1238, 1280*

brain **6**:*795*
  cartilaginous **3**:*436;*
      4:*460,* 461; **8**:*1060,*
      1096; **9**:*1238, 1280*
  elopomorph **3**:*436*
  evolution of tetrapods
      from **9**:1158–1159
  fastest **8**:1064
  fins **9**:1174
  flatfish **9**:1244
  gas exchange by
      8:*1012*
  hormones
      calcium-regulating
          2:268
      growth-inhibiting
          2:267
  intestines **2**:181
  jawed **1**:*122*–123;
      4:*446;* **8**:1097
  jawless (agnathans)
      3:*436;* **4**:*446–447,*
      460; **8**:*1060, 1096;*
      9:*1238, 1280*
  lobe-finned **1**:*122,*
      123; **3**:*436;* **8**:*1060,*
      1096; **9**:*1158,*
      1174, *1280*
  with lungs **8**:*1020*
  ray-finned **3**:*436;*
      8:*1060,* 1061,
      1096, 1097; **9**:1174,
      1280, 1281
  scales **4**:*463;* **9**:*1283*
  swim muscles
      6:764–765
  teeth **2**:174
flounder **9**:1241

G

gar, scales **9**:1283
gnathostomes
      1:*122*–123; **3**:*436;*
      4:*460;* **8**:*1060,*
      1096, 1097;
      9:*1238, 1280*
grayling **9**:*1280*

guitarfish **9**:*1238,*
      1243
gulper eel **3:436–445**
  circulatory and
      respiratory systems
      3:437, *443*
  common Atlantic
      3:*437, 438, 441,*
      444, 445
  digestive and excretory
      systems **3**:437, *444*
  external anatomy
      3:437, *438–440*
  nervous system **3**:437,
      442
  reproductive system
      3:437, *445*
  skeletal and muscular
      systems **3**:437,
      441

H

hagfish **4:446–459**
  Atlantic **4**:*446,* 449,
      452, 453, 455, *456,*
      457, 459
  circulatory and
      respiratory systems
      1:111, *112;* **4**:447,
      455–456
  digestive and excretory
      systems **4**:447,
      457–458
  external anatomy
      4:447, *449–451*
  nervous system **4**:447,
      453–454
  reproductive system
      4:447, *459*
  skeletal and muscular
      systems **4**:447, *452;*
      9:*1159*
hammerhead shark
      4:460–475
  circulatory and
      respiratory systems
      4:461, *472–473*

digestive and excretory
    systems 4:461, *474*
external anatomy
    4:461, *462–463*
gills 4:*472–473*
great 4:*461*
muscular system 4:461,
    466–467
nervous system 4:461,
    468–471
    electroreceptors
        6:*787*
reproductive system
    4:461, *475*; 7:*997*
scalloped 4:*462*
skeletal system 4:461,
    464–465
hatchetfish 3:440
houndshark 4:*460*

I

icefish 8:*1025*

L

labyrinthodonts 1:126
lamprey 4:*446–4478*
    hemoglobin 4:456
    mouth 4:*458*
    vertebrae 9:*1172*
lanternfish 3:439, 440
lionfish, venom 2:275
lungfish 1:123; 8:*1020*
    African 1:*122*, 129
    Australian 1:*122*, 129;
        8:*1020*
    South American 1:*122*,
        129

M

mackerel 3:443; 8:1027
    swim muscles
        6:*764–765*
marlin 8:1064, *1065*
    black, migration
        8:1073
mullet 8:*1060*

P

perch 8:*1060*, 1061;
    9:*1280*
    scales 9:*1283*
pike 9:*1280*
pipefish 8:*1096*, 1098,
    1109
pufferfish, venom
    6:783

R

ray 4:461; 9:*1238*, 1239
    blue-spotted 9:*1245*
    buoyancy 9:*1249*
    butterfly 9:*1251*
    cow-nose 9:*1244*
    eagle 9:1239, 1244,
        1250, 1251
    electric/torpedo
        9:*1238*
    electroreceptors 6:*787*
    manta/devil 9:*1244*,
        1249
    reproduction 9:1251
    scales 9:*1283*
    spotted eagle 9:*1241*
    thornback 9:*1250*

S

sailfish **8:1060–1073**
    Atlantic 8:*1060*, *1066*
    circulatory and
        respiratory systems
        8:1061, *1070–1071*
    digestive and excretory
        systems 8:1061,
        1072
    external anatomy
        8:1061, *1062–1064*
    Indo-Pacific 8:*1060*
    muscular system
        8:1061, *1066–1067*
    nervous system 8:1061,
        1068–1069
    reproductive system
        8:1061, *1073*

skeletal system 8:1061,
    1065
    swim bladder 8:1072
salmon 8:1066, 1067;
    9:*1280*, 1281
    Atlantic 9:*1280*, 1281
    Pacific 9:1281, 1293
sarcopterygians
    9:*1158–1159*
sawfish 9:*1238*
scombroids 8:1061
sea dragon 8:*1100*
sea horse **8:1096–1109**
    camouflage 8:1100
    circulatory and
        respiratory systems
        8:1097, *1104–1105*
    clicking noises 8:1103
    digestive and excretory
        systems 8:1097,
        1106–1107
    evolution 8:1100
    external anatomy
        8:1097, *1099–1100*
    eyes and vision 8:1100,
        1103
    gills 8:1100, *1105*
    golden 8:*1102*
    long-snout 8:*1107*
    nervous system 8:1097,
        1103
    number of species
        8:1098
    as predators and prey
        8:1107
    pygmy 8:1098, *1107*
    reproductive system
        8:1097, *1108–1109*
    skeletal system 8:1097,
        1101–1102
    snout of 8:*1107*
    spotted 8:*1097, 1099,
        1101, 1103, 1104,
        1105, 1106, 1109*
    thorny 8:*1098*
    tiger tail 8:*1105*
    White's 8:*1102*

shark 4:461; 9:*1238*
    angel 4:*463*; 9:*1242*
        spiracles 4:*473*
    buoyancy 9:1249
    cat 4:463
    electroreceptors 6:*787*
    frilled 4:*463*
    great white 4:*467*, 469,
        472; 8:*1027*
    ground 4:*460*, 461
    hammerhead *see*
        hammerhead shark
    requiem 4:*460*
    respiratory system,
        ventilating gills
        8:*1027*
    saw 9:*1242*
    scales 9:*1283*
    skeletal system 4:461,
        464–465; 9:*1169*
    vertebrae 9:*1172*
    weasel 4:*460*
skate 9:*1238*, 1239
    electroreceptors 9:*1247*
    reproduction 9:1251
snake mackerel 8:*1060*
spearfish 8:*1060*, 1064
    Mediterranean 8:1064
stickleback 8:1100
stingray **9:1238–1251**
    blue-nose 9:1244
    blue-spotted 9:1244
    blunt-nose 9:1244
    circulatory and
        respiratory systems
        9:1239, *1248*
    diamond 9:1250,
        1251
    digestive and excretory
        systems 9:1239,
        1249–1250
    external anatomy
        9:1239, *1240–1241*
    nervous system 9:1239,
        1245–1247
    reproductive system
        9:1239, *1251*

stingray *(continued)*
round 9:*1247*
skeletal and muscular
systems 9:1239,
1242–1244
southern 9:*1239,
1240,* 1241, 1244,
1251
venomous sting
9:*1241*
yellow 9:*1250*
swallower 3:*436*
swordfish 8:*1060,* 1064

T

tapirfish 3:*436*
tarpon 3:*436*
toadfish, oyster 6:765
trout 8:1066, 1067;
9:1280–1293
brook 9:*1280*
brown 9:*1281, 1284*
circulatory and
respiratory systems
9:1281, *1289*
cutthroat 9:*1292*
digestive and excretory
systems 9:1281,
1290–1291
external anatomy
9:1281, *1282–1284*
gila 9:*1292*
interbreeding 9:1292
nervous system
9:1281, *1287–1288*
lateral line 6:*790;*
9:*1287, 1288*
sense of smell
9:*1284,* 1287–1288
rainbow 9:*1280, 1282,
1283, 1285, 1286,
1290–1293*
reproductive system
9:1281, *1292*
skeletal and muscular
systems 9:1281,
1285–1286

tuna 8:*1060,* 1061
body temperature
1:120
gills 8:1027, *1071*
muscles and swimming
6:764–765;
8:*1066–1067*
yellowfin 8:*1067*
turbot 9:1241

W

whipray, mangrove
9:1244
whitefish, northern
9:1281

Invertebrates

A

alderfly 4:*538*
ammonite 6:*815*
anemone
coral-like 8:*1084*
tube 8:*1084*
see also sea anemone
anisozygopterans 2:*208,*
209
annelid worms 2:234
hormones 2:269
mouthparts 2:236
skeletal system
9:1163–1164
anomurans 1:*132,* 133
ant 1:**26–35;** 4:*526, 538*
anatomy external 1:27,
28–29
internal 1:27, *30*
black garden 1:*29*
carpenter 1:*28,* 30, *34*
castes 1:29
chemical
communication
1:31
differs from termites
1:29
digestive and excretory
systems 1:27, *32–33*

dracula 1:*26,* 27, 33
as gardeners or farmers
1:*33*
harvester 1:*26,* 27, 32
honeypot 1:*26,* 27, 33
king and queen 1:35
leaf-cutter 1:*26,* 27, *33,*
35; 6:773
nervous system 1:27,
31
ovipositor function
1:35
reproductive system
1:27, *34–35*
respiration 1:30
soldier 1:29
trap-jaw 1:*26,* 27, 28
anthozoans 5:*636,* 638,
639; 8:*1084*
aphid 1:33; 5:*692*
apocritans 4:527
arachnids 8:*1074;*
9:*1252,* 1253
skeletal system
9:1166–1167
araneomorphs 9:*1252*
arthropods 1:*26, 132;*
2:208; 3:*332;*
4:*494, 526, 538;*
5:*684, 692;* 8:*1074;*
9:*1252*
digestive system 2:179
exoskeleton 9:1165,
1166–1167
molting 2:267
mouthparts 2:175
nervous systems 6:793

B

backswimmer
8:1017–1018, 1025
barnacle 3:*378*
bee 1:*26,* 27; 4:*494, 526,
538*
Africanized (killer)
4:530–531
carpenter 4:*526,* 531

euglossine/orchid
6:833
leafcutter 4:*526*
long-tongued 4:*526,*
527
plasterer 4:*526*
stingless 4:*526,* 527
sweat 4:*526*
wings 2:211
see also honeybee
beetle 4:*494, 538;* 5:*692;*
10:*1344*
"ancient" 10:*1344*
antennae 10:*1347,*
1348
bombadier, toxins
2:275
click 10:1349, *1350*
cucujo 10:1354
evolution 10:1348
external anatomy
10:1346–1348
goliath 8:*1021*
ground 10:*1344*
longhorn 8:1148
metamorphosis
10:1358
rove 10:1347–1348
skiff 10:*1344*
wings 2:211
biddy 2:*208*
bivalves 3:*342*
catch muscles 6:768
circulatory system
1:107
heterodont 3:*342,* 343
nervous system 3:*353*
blood fluke, asexual
reproduction
7:987
bloodworm 4:547
blowfly 4:*538,* 539
sheep, maggots 4:548
botfly 4:539, 547; 8:*1021*
sheep, ovoviviparity
4:549
brachyurans 1:*132,* 133

brittle star 9:*1224*, 1225, 1228
bumblebee 4:*526*, 527
butterfly 4:*494*, 495, *538*, 541; 5:*692*
  antennae 4:496
  wings 2:211

C

caddisfly 4:*538*
cancridans 1:*132*
capsid bug 10:1304
caterpillars 4:504–*505*
  digestive system 2:*179*
  metamorphosis 2:267
  moth 4:*499*, 501, *505*
  nervous system 4:502
centipede 3:*332*, 333; 8:*1074*
  giant *see* giant centipede
  house 3:337, 338
  soil 3:*332*
  stone 3:*332*
cephalopods 6:*812*, 813
  color change 6:816
chelicerates 9:*1252*–1253
clam
  catch muscles 6:768
  dipper 3:*342*, 353
  giant *see* giant clam
  nut 3:352
  razor 3:*342*, 343, 351
  soft-shell 3:*342*
  vent 3:*342*, 343, 352
clitellate worms 2:*234*, 235
clubtail 2:*208*
cnidarians 5:*636*, 637; 8:*1084*, 1085, 1087
  body wall 5:*642*
  nerve cells 8:1090
  venoms used by 5:640
cockle 3:*342*, 343, 345, 347–348, 350
coleoids 6:*812*, 813

conch, queen 9:*1164*
copepod, filter feeding 2:175
coral 8:1085, 1087
  black 8:*1084*
  crown-of-thorns starfish and 9:1235
  soft 8:1087
  stony 8:*1084*, 1087
crab 1:**132–141**; 5:*684*
  Atlantic rock 1:*132*, 133, 141
  blue 1:136
  boxer 1:135
  circulatory and respiratory systems 1:133, *138*
  coral 1:*141*
  digestive and excretory systems 1:133, *139*
  Dungeness 1:*132*, 133
  edible 1:*132*, *134*, *136*, *137*, *138*, *139*, *140*
  external anatomy 1:133, *134–135*
  fiddler 1:135, 140
  green 1:*136*
  hairy rock 1:*132*
  hermit 1:*133*, 135; 8:1087
  horseshoe 8:1017; 9:*1252*
  immunity 5:611
  land 1:*132*
  mole 1:133
  molting 1:141
  muscular system 1:133, *136*; 6:762, 763
  nervous system 1:133, *137*
  Pacific rock 1:*132*
  pea 3:347
  reproductive system 1:133, *140–141*
  robber 1:136
  sand 1:133
  spider 1:*132*

swimming crabs 1:*132*
  true 1:133
  walking sideways 1:*136*
crayfish 1:133
  claws 2:175
  muscles 6:762
  heart 6:767
cricket
  excretion 2:*187*
  sound detection 6:790
crinoids 9:*1224*
crustaceans 1:*132*–133; 3:332; 5:*684*–685; 8:*1074*; 9:*1252*
  digestive system 2:179
  excretory system 2:187
  exoskeleton 9:1163
  hormones 2:269
ctenophores 5:636–637
cubozoans 5:638, 639
cuttlefish 6:*812*, 813
  color change 6:*816*
  ink 6:817
  large nerve cells 6:820
cyamid (whale louse) 3:*378*

D

damselfly 2:*208*, 209; 10:*1312*
  nymphs, breathing underwater 2:216; 8:1017
darner 2:*208*, 209
decapods 1:*132*, 133; 5:*684*, 685; 6:*812*
  claw muscles 6:762
  neurogenic hearts 6:767
digestive systems 2:178–179
  intestines 2:183
  mouthparts 2:175
dragonfly 2:**208–219**; 4:528; 5:*692*
  brain 2:215

circulatory and respiratory system 2:209, 213, 216–217
  digestive system 2:214
  emperor 2:*208*, 209
    circulatory system 2:*216*
    external anatomy 2:*210*
    internal anatomy 2:*214*
    nervous system 2:*215*
    reproduction 7:*996*
  external anatomy 2:209, 210–213
  eyes and vision 2:212
  flight 2:212, 214
  hunting by 2:212–213
  internal anatomy 2:209, 214
  molting 2:*211*, 213
  muscles 2:214
  nervous system 2:209, 215
  nymphs 2:218–219
    anatomy 2:213
    breathing underwater 2:216, *217*; 8:1017
    mouthparts 2:*219*
    movement 2:216, *217*
  rectum 2:183, 216
  reproductive system 2:209, 213, 218–219
  wings 2:211–212, 214
dung fly 4:548

E

earthworm 2:**234–243**
  brain 2:240
  breaking in two and regeneration 2:239
  circulatory and respiratory systems 1:107; 2:235, 241

earthworm *(continued)*
  digestive system **2**:183, 184, 235, 238, 242
  excretory system **2**:235, 237, 238, 242
  external anatomy **2**:235, 236–237
  giant **2**:*234*
  internal anatomy **2**:235, 238–239
  movement **2**:237, 239; **6**:762
  muscles **2**:238; **6**:762
  nervous system **2**:235, 240; **6**:792–*793*
  reproductive system **2**:235, 238, 243
  response to attack **2**:240
  sense organs **2**:240
  skeletal system **2**:239; **9**:*1163*–1164
  skin **2**:236, *237*
echinoderms **9**:*1224*–1225
  body shape **9**:1228
  gas exchange by **8**:*1012*
  nervous systems **6**:793
  related to chordates? **9**:1237
  skeletal systems **9**:1164
endopterygotes **1**:*26*; **4**:495, 526, 538–539
eurypterids **8**:1077
excretory systems **2**:187

F

fan worm **8**:*1027*
feather star **9**:1225, 1228
firefly **10**:*1344*, 1348, 1349, 1350–1351
flatworm
  diffusion of oxygen **8**:*1029*

excretion **2**:175, *187*
gut **2**:183
nervous system **6**:*781*, 792
flea **4**:*538*
flesh fly *see* fly, flesh
fluke, blood *see* blood fluke
fly **4**:*538*, 539; **5**:*692*
  aschizan **4**:*538*
  brachyceran **4**:*538*
  cyclorrhaphan **4**:*538*, 540, 546
  dung **4**:548
  flesh **4**:*538*, 539
  fruit (vinegar) **4**:*539*
  head **4**:*546*
    mutations **6**:793
  hover **4**:530, *538*, 539, 547; **6**:762
  humpback **4**:*538*
  nematoceran **4**:*538*
  parasitic **4**:*538*
  schizophoran **4**:*538*
  stable **4**:*539*
  tsetse **4**:*546*, 548, 549
  warble **4**:539, 547
fruit fly *see* fly, fruit

G

gastropods, circulatory system **1**:107
geoduck **3**:*342*, 345, *347*
giant centipede **3**:*332–341*
  anatomy
    external **3**:333, *334–335*
    internal **3**:333, *336*
  circulatory and respiratory systems **3**:333, *338*
  digestive and excretory systems **3**:333, *339–340*
  giant desert centipede **3**:333

giant redheaded centipede **3**:*334*, *335, 336, 337, 339, 341*
  nervous system **3**:333
  reproductive system **3**:333, *341*
giant clam **3**:*342–355*
  anatomy
    external **3**:343, *344–345*
    internal **3**:343, *346–347*
  circulatory and respiratory systems **3**:343, *354*; **8**:*1016*
  digestive and excretory systems **3**:343, *352*
  muscular system **3**:343, *350–351*
  nervous system **3**:343, *353*
  reproductive system **3**:343, *355*
glowworm **10**:1348, 1349, 1350
grasshopper, nervous system **6**:*781*
grayback **2**:*208*

H

harvestman **8**:*1074*, 1075
hawkmoth **4**:*494–505*
  anatomy
    external **4**:495, *496–499*
    internal **4**:495, *500–501*
  circulatory and respiratory systems **4**:495, *503*
  death's-head **4**:499
  hummingbird **4**:*498*
  nervous system **4**:495
  poplar **4**:*495*
  reproductive system **4**:495, *504–505*

hexapods **1**:*26*; **2**:208; **3**:*332*; **4**:494–495, 526, 538; **5**:*692*; **10**:*1344*
honeybee **1**:*26*; **4**:*526–537*
  anatomy
    external **4**:527, *528–531*
    internal **4**:527, *532–533*
  Asian dwarf **4**:*526*
  Borneo **4**:*526*
  circulatory and respiratory systems **1**:*117*; **4**:527, *535*
  digestive and excretory systems **4**:527, *536*
  dwarf **4**:*526*
  eastern **4**:*526*
  giant **4**:*526*
  Himalayan **4**:*526*
  nervous system **4**:*534*
  reproductive system **4**:527, *537*
hornworm, tobacco **4**:497
housefly **4**:*538–549*
  anatomy
    external **4**:539, *540–543*
    internal **4**:539, *544*
  circulatory and respiratory systems **4**:539, *547*
  digestive and excretory systems **4**:539, *548*
  feet **4**:*542*
  halteres **2**:211; **4**:*540*, 542
  maggots (larvae) **4**:547, 548, *549*
  nervous system **4**:539, *545–546*
  reproductive system **4**:539, *549*
  sucking liquids **4**:*541*

hover fly *see* fly, hover
hydra **1**:106; **2**:183
  nervous system **6**:*781*
  reproduction, asexual
      7:*987–988*
hydrozoans **5**:*636, 637,*
      638, 639, 640;
      7:986, 987; **8**:*1084*
hymenopterans **4**:*526*
  narrow-waisted **4**:*526*

I

insects **1**:*26;* **2**:208–209;
      4:495, 526, *526,*
      538; **5**:*692;* **8**:*1074;*
      10:*1344,* 1345
  circulatory system
      1:107
  digestive system **2**:179,
      183
  evolution **10**:1348
  excretion by **2**:186
  hormones **2**:267, 269
  immunity **5**:608
  largest **2**:217; **8**:*1021*
  molting **2**:267
  skeletal system
      9:*1166–1167*
  wing movement
      6:*762*

J

jellyfish **5**:**636–647**
  anatomy
      external **5**:637,
          639–640
      internal **5**:637,
          641–642
  bell **5**:*638*
  box **5**:*636,* 638, 640,
      643; **8**:*1084*
  common moon **5**:*636,*
      638, *639, 641,* 642,
      643, *644,* 646–647
  digestive and excretory
      systems **5**:637,
      644–645

dunce-cap **5**:*643*
  lacking blood **3**:354
  lion's mane **5**:*637,* 638,
      640
  moon **5**:*636,* 638, *639,*
      641, 643, *644,* 646,
      646–647; **7**:*987*
  nervous system **3**:353;
      5:637, *641, 643;*
      6:*789*
  purple stinger **5**:*645*
  reproductive system
      5:637, *646–647*
  skeletal system **5**:*642;*
      9:*1162*
  true **5**:*636;* **8**:*1084*

K

krill **1**:16, 133

L

lacewing **4**:*538*
ladybug **10**:*1344, 1348,*
      1356
lamp-mussel **3**:355
lancelet **4**:454
  hepatic cecum **2**:184
  hormones **2**:269
leech **2**:*234,* 235
  mouthparts **2**:236
lepidopterans **4**:495, 496,
      504
lobster **1**:133; **5**:**684–691**
  American **5**:*684,* 685,
      686, 687, 689, 691
  anatomy
      external **1**:135; **5**:685,
          686–687; **9**:*1167*
      internal **5**:685,
          688–689
  clawed **5**:685
      claws **2**:175; **5**:*687*
      muscles **6**:762
  heart **6**:767
  nervous system **5**:685,
      690
  pincer **5**:*684*

reproductive system
      5:685, *691*
  rock **5**:*684*
  slipper **5**:*684*
  spiny **5**:*684*
  true **5**:*684*
louse **5**:**692–699**
  anatomy
      external **5**:693,
          694–695
      internal **5**:693, *696*
  digestive and excretory
      systems **5**:693,
      697–698
  human body **5**:693,
      698, 699
  human head **5**:*693,*
      694, 699
  pubic (crab) **5**:693, *695*
  reproductive system
      5:693, *699*
  shaft **5**:693
  true lice **5**:*692*
  whale (cyamid) **3**:*378*
lugworm **8**:1026
lumbricid worms **2**:*234*

M

maggots, rat-tailed **4**:547
malacostracans **1**:*132,*
      133; **5**:*684,* 685
mayfly, nymphs **8**:1017
metazoans **8**:*1084*–1085;
      9:1224
midge **4**:548
  larvae **4**:547
millipede **3**:*332,* 333;
      8:*1074*
  defenses **3**:*340*
  diplosegments **3**:335
  exoskeleton **9**:1166
  polydesmid **3**:340
  porcupine–Velcro
      3:*340*
mite **8**:*1074,* 1075;
      9:*1252*
  on bees **4**:*531*

mollusks **3**:*342;*
      6:*812*–813
  digestive system **2**:179
  excretory system **2**:187
  radula **2**:175, 179
  shells **9**:*1164*–1165
mosquito
  larvae **4**:547
  mouthparts **4**:*541*
  sucking blood **4**:546
moth **4**:*494,* 495, *538,*
      541
  apatelodod **4**:*494*
  bombycoid **4**:*494,* 495
  giant atlas **4**:495
  lappet **4**:*494*
  silkworm **4**:*494,* 504
  tent caterpillar **4**:*494*
  tobacco hornworm
      4:*496*
  vampire **4**:497
  see also hawkmoth
mussel **3**:*342,* 350
  catch muscles **6**:768
  freshwater **3**:*342,* 355
mygalomorphs **9**:*1252*
myriapods **3**:*332*–333

N

nautiloids **6**:*812,* 813
  buoyancy **6**:815
nematode (roundworm)
      6:*774*
  nervous system **6**:781,
      *792–793*dipterans
      4:539
night crawler **2**:*234,* 235
  digestive tract **2**:*242*
  external anatomy
      2:*236*
  gaseous exchange
      2:*241*
  internal anatomy
      2:*238, 239*
  nervous system **2**:240
  reproductive system
      2:*243*

O

octopods **6**:*812*
octopus **6**:**812–823**
  blue-ringed **6**:822
  circulatory and
      respiratory systems
      6:813, *821*
  common **6**:*815–816,
      819, 821, 822,* 823
    mouth **2**:*175*
  digestive and excretory
      systems **6**:813, *822*
  external anatomy
      6:813, 814–*817*
  eye **6**:*788*, 817, *820*
  giant **6**:*818*
  ink **6**:*817*
  jet propulsion **6**:*818*
  muscular system **6**:813,
      818
  nervous system **6**:792,
      813, *819–820*
  reproductive system
      6:813, *823*
  tentacles **9**:*1165*
oligochaete worms
    2:234
oyster **3**:*342,* 345, 355
  American **3**:349

P

palolo worms **7**:988
pauropods **3**:*332,* 333
  legs **3**:*335*
pill bug (wood louse)
    5:692
planarians, excretion
    2:*187*
polychaete worms
    2:234, *235*
  gills **2**:241
  mouthparts **2**:236
  and reproduction **7**:988
Portuguese man-of-war
    5:637, *640*
pseudoscorpion **8**:*1074,*
    1075

R

ragworm **9**:1163
ram's horn shells **6**:813
rat-tailed maggot **8**:*1017*
roundworm (nematode)
    6:*774*

S

salamander, nervous
    system **6**:*781*
sand dollar **9**:*1224,*
    1225
sandworm, jaws **2**:*178*
sawfly **1**:*26,* 27; **4**:*526*
scallop **3**:*342,* 345, 353
  escape response **9**:1234
  eyes **3**:*353*
  jet propulsion **3**:*351*
scarab **10**:*1344*
scorpion **8**:**1074–1083**;
    9:*1252*
  ancestors **8**:1077
  black **8**:*1075*
  circulatory and
      respiratory systems
      8:*1022,* 1075, *1080*
  digestive and excretory
      systems **8**:1075,
      1081
  emperor **8**:*1076, 1077,
      1079–1083*
  external anatomy
      8:1075, *1076–1077*
  muscular system
      8:1075, *1078*
  nervous system **8**:1075,
      1079
  reproductive system
      8:1075, *1082–1083*
  sting and venom
      8:*1076,* 1081
  see also wind scorpion
scorpionfly **4**:*538*
sea anemone
    8:**1084–1095**
  burrowing **8**:*1091,
      1094*

  common plumose
      8:1085, *1086,
      1087, 1088, 1095;*
      9:*1166*
  crabs' use of **1**:135;
      8:1087
  digestive and excretory
      systems **8**:1085,
      1092–1093
  diploblastic layers
      8:1089
  external anatomy
      8:1085, *1086–1087*
  giant **8**:1086
  internal anatomy
      8:1085,
      1088–1089; **9**:*1166*
  lacking blood **3**:354
  larvae **8**:*1094*
  muscles **6**:*755*
  nervous system **8**:1085,
      1090
  northeast Pacific
      8:1095
  plumose **8**:1085
  reproductive system
      8:1085, *1094–1095*
  respiratory system
      8:1085, *1091*
  tentacles **8**:1086, *1087*
  defense **8**:1090
sea cucumber **9**:*1224,*
    1225, 1228
sea daisy **9**:*1224,* 1225
sea fan **8**:1085, 1087
sea lily **9**:1225, 1228
sea pen **8**:1087
sea slug, and anemone
    stinging cells **8**:1092
sea spider **9**:*1252*
sea squirt, pharynx **2**:*176*
sea star
  sunflower **9**:1225, 1234
  see also starfish
sea urchin **9**:*1224,* 1225,
    1228
  growth **9**:1167

  sea otters and **6**:859
  skeletal system **9**:*1164,*
      1167
  slate pencil urchin
      9:*1228*
sea wasp **5**:638
shipworm **3**:*342*
shrimp **1**:133; **5**:*684*
skimmer **2**:*208*
skipper **4**:*494*
slug **3**:342
snail **3**:342
  digestive system **2**:*179*
  lung **8**:*1019*
snakefly **4**:*538*
solpugids **8**:1075
spider **8**:*1074;* **9**:*1252,*
    1253
  crab **9**:1256
  digestive system **2**:*179*
  feeding **2**:175
  funnel-web **9**:1263
  giant trapdoor **9**:*1252*
  jumping **9**:1260
  mechanoreceptors
      6:*791;* **9**:*1259–1260*
  mouthparts **2**:175
  respiratory system
      8:*1022*
  water **8**:*1018*
  world's largest **9**:*1253*
  see also tarantula
sponge
  skeletal system **9**:1162
  ventilation **8**:*1026*
springtail **4**:494–495
squid **3**:*345;* **6**:*812*
  circulatory system
      1:107
  color change **6**:816
  dumpling **6**:813
  eyes **6**:788
  ink **6**:817
  nervous system
      giant nerve cells
          6:783, *820*
  vampire **6**:*812*

stable fly 4:539
starfish 9:1224–1237
  circulatory and
    respiratory systems
    9:1225, 1233
  crown-of-thorns
    9:1224, 1235
  digestive and excretory
    systems 9:1225,
    1234–1235
  external anatomy
    9:1225, 1226–1228
  internal anatomy
    9:1225, 1229–1231
  nervous system 9:1225,
    1232
  reproductive system
    9:1225, 1236–1237
  skeletal system 9:1164
  sunflower 9:1224, 1226
  see also sea star
stonefly, nymphs 8:1017
strepsipterans 4:542
  halteres 2:211
sun scorpion see wind
  scorpion
sun star, common
  9:1224
symphylans 3:332, 333,
  341

T
tapeworm 2:182, 183
tarantula 8:1022;
  9:1252–1265
  circulatory and
    respiratory systems
    9:1253, 1261
  digestive and excretory
    systems 9:1253,
    1262–1263
  external anatomy
    9:1253, 1254–1256
  goliath 9:1253, 1254,
    1261
  internal anatomy
    9:1253, 1257

king baboon 9:1263
Mexican red-kneed
  9:1260
nervous system 9:1253,
  1259–1260
reproductive system
  9:1253, 1264–1265
skeletal and muscular
  systems 9:1253,
  1258
termite 1:26, 29
  king and queen 1:35
tick 5:698; 8:1074–1075;
  9:1252
trilobite, exoskeleton
  9:1166
tsetse fly 4:546, 548, 549
tunicate 3:343; 8:1025
twisted-winged parasites
  4:538, 542

U
uniramians 2:208; 3:332;
  4:494; 9:1252

W
warble fly see fly, warble
wasp 1:26, 27; 4:547
  parasitic 1:35; 4:526
water flea, alternation of
  generations 7:989
water scorpion 8:1017
wedge shell 3:342, 343,
  347–348, 351
weevil 4:528;
  10:1344–1359
  bearded 10:1351
  black vine 10:1359
  chestnut 10:1349,
    1359
  circulatory and
    respiratory systems
    10:1345,
    1355–1356
  cotton-boll 10:1346,
    1352, 1354, 1355,
    1357

cycad 10:1344
digestive and excretory
  systems 10:1345
external anatomy
  10:1345,
  1346–1351
fossil 10:1348
fungus 10:1344
hazlenut 10:1359
internal anatomy
  10:1345,
  1352–1353
nervous system
  10:1345, 1353,
  1354
nut 10:1358–1359
pear-shaped 10:1344
pine 10:1353
reproductive system
  10:1345,
  1357–1359
thief 10:1344
wind scorpion (sun
  scorpion) 8:1074,
  1075, 1078
wood louse (pill bug)
  5:692
  exoskeleton 9:1166

Y
yellow jacket 4:527

Mammals

A
aardvark
  brain 3:328
  sense of smell 3:328
  teeth 3:321
aardwolf 5:592, 593, 598,
  599, 603, 604
  fur markings
    5:594–595
  skull 5:597
  teeth 5:597
  toes 5:594

tongue 5:599
anteater 3:312, 313;
  9:1176, 1177
  digestive system 9:1191
  giant see giant anteater
  silky 3:312, 313, 315
    claws 3:316
    skull 3:319
    tail 3:317, 324
    teeth 3:321
  spiny see echidna
antelope 3:356, 357;
  7:964, 967–968;
  10:1360
  dwarf 10:1360
  four-horned 10:1365,
    1367, 1376
  grazing 10:1360
  limbs 10:1369, 1371
  spiral-horned 10:1360,
    1364
  toes 10:1366, 1367
ape 1:88, 89; 4:550;
  6:724
  great 1:88, 89
armadillo 3:312, 313;
  9:1176, 1177, 1181
  brain 3:328
  claws 3:316
  exoskeleton 9:1169
  giant 3:321; 9:1185
  nine-banded 9:1169
  sense of smell 3:328
  teeth 3:321
  yellow 9:1181
artiodactyls see ungulates,
  even-toed
ass 8:1032
  African 10:1418
  Asiatic 10:1418
  horse-ass hybrids
    10:1437

B
babirusa 4:506
baboon 6:724, 725
  hands and feet 6:728

badger **10**:*1324*, 1333
 American **10**:1331,
  1333
 European/Eurasian
  10:1331, 1333,
  1339
bat **3**:292
 blossom **3**:*292*
 circulatory system
  1:*106, 115*
 fruit *see* fruit bat
 lesser long-nosed
  8:*1058*
 long-tongued **7**:*1002*
 pollination by **7**:*1002*;
  8:*1058*
 vampire, teeth **2**:*174*
 wings **9**:*1175*
bear **3**:*410*, 411; **5**:*592,
  664*; **6**:*844*; **7**:*928*;
  10:*1324*
 American black **3**:*410*,
  411, *415*
 Asian black **3**:*410*, 411,
  414
 brown **3**:*410*, 411, *415*
 dawn **3**:415
 grizzly *see* grizzly bear
 Kodiak **3**:411
 polar **3**:*410*, 411, *415*
  hibernation **3**:*425*
  skeleton **3**:*418*
  skull **3**:*417*
 short-faced **3**:*415*
 sloth **3**:*410*, *414*
 skull **3**:*419*
 spectacled **3**:*410*, *418*
 skeleton **3**:*418*
 sun **3**:*410*, *414*
beaver, tail **7**:955
beluga **2**:*188*, 190
 communication **2**:193
 facial expressions
  2:193
bettong **5**:649, 656
blackfish **2**:*188*
bobcat **7**:945

bonobo *see* chimpanzee,
  pygmy
boutu (Amazonian river
  dolphin) **2**:*195*, 205
bovids
 evolution **10**:1377
 horns **10**:1364–*1365*,
  1369
 skull **10**:1369
bushbuck **10**:*1376*
 skull **10**:*1369*

C

calcium regulation **2**:268
camel **3**:*356*; **4**:*506*;
  7:*964*; **10**:*1360*
 stomach **10**:1377
capybara **7**:*898*
caribou (reindeer) **7**:*964*,
  966, *972*, 974
 females with antlers
  7:*982*
carnivores **3**:*410*; **5**:*592*,
  664, 665; **6**:*844*,
  845; **7**:*928*, 929;
  8:*1110*–1111;
  10:*1324*, 1325,
  1382, 1383
 catlike **5**:*592, 664*;
  10:*1324*
 digestive system
  8:1122
 intestines **2**:181;
  8:1122–1123
 land **3**:*410*; **10**:*1382*
 teeth **2**:174; **5**:592
cat **5**:*592, 664*, 665;
  6:*844*; **7**:*928*, 929
 big cats **5**:*664*, 665;
  7:*928*, 932
  eyes **7**:939
 classification **7**:932
 coat patterns **7**:*931*
 Iriomote **7**:932
 Manx **7**:935
 small cats **5**:*664*, 668;
  7:*928*, 929, 932

eyes **7**:*939*
 hyoid apparatus
  7:*935*
cattle **3**:*356*, 357; **7**:*964*;
  10:*1360*
 stomach **5**:*659*
 wild **10**:*1360*
cavy **7**:*898*
cebids **6**:*729*, 730
cercopithecines **6**:725
cetaceans **2**:189;
  3:374–375
 limb morphology
  2:192, 196
 skin **2**:191
cheetah **5**:665, *673, 678*;
  7:*932*
chickaree **8**:*1148*
chimpanzee **1**:88–105;
  4:*550*
 ancestral diets **1**:101
 babies **1**:*104–105*
 bipedalism **1**:*94*, 97
 circulatory and
  respiratory systems
  1:89, *100*
 common **1**:89, *90, 91,
  92, 93, 95, 98, 100,
  101, 102, 103*
 digestive and excretory
  systems **1**:89,
  101–102
 external anatomy **1**:89,
  90–91
 facial expressions **1**:*96*
 language skills **1**:99
 meat eaters **1**:102
 muscular system **1**:89,
  95–97
 nervous system **1**:89,
  98–99
 opposable thumbs and
  toes **1**:*97*
 pygmy (bonobo) **1**:89,
  91, *94*
 gestation **1**:103
 promiscuity **4**:573

reproductive system
  1:89, *103–105*
 skeletal system **1**:89,
  92–94
 social structure **1**:105
 strength **1**:97
 teeth **4**:569
 tool use **1**:97, 104
 vocalizations **1**:100
chinchilla **7**:*898*
chipmunk **9**:*1208*
civet **5**:*592, 664*; **7**:*928*;
  10:*1324*
cloaca **2**:185
colobines **6**:725, 739
colobus *see* monkey,
  colobus
cottontail **4**:*476*
coyote **10**:*1382*

D

deer **3**:*356*, 357; **7**:*964*,
  966, 967, 968–969;
  10:*1360*
 antlers **7**:*965, 966, 969,
  971–972, 974*, 975,
  980, *982*
 brocket **7**:981
 Chinese water **7**:969,
  983
 chronic wasting disease
  (CWD) **7**:975
 crossbreeds **7**:981
 "endurance hunting"
  of **7**:974
 fallow **7**:*964, 966, 974,
  983*
 lesser Malay mouse
  7:*969*
 mouse **3**:*356*
 mule **7**:*964*, 982
 muntjac **7**:*964*, 969,
  981
 musculature
  7:*973–974*
 musk **3**:*356*; **7**:*964*;
  10:*1360*

names for males,
females and young
7:983
pampas **7**:*969*
red *see* red deer
red brocket **7**:*969*
roe **7**:*964*, 966,
982
sika **7**:981
skeleton **7**:970
with white rumps
7:967
white-tailed **7**:*964*,
966, *980*, 982
dik-dik, territory-
marking **2**:274
dingo **7**:885; **10**:*1382*
dog **3**:*410*; **5**:*592, 664*;
6:*844*; **7**:*928*;
10:*1324, 1382*,
1384
detecting seizures
10:1395
domestic **10**:*1388*
sniffer **10**:1396
tail-wagging
10:1393
true **10**:*1382*, 1384
dolphin **2**:**188–207**;
4:*506*
beak or snout **2**:192,
195
blowhole **2**:195, *196*,
203
blubber **2**:191, 198
bottle-nosed **2**:*190*,
191
brain **2**:201
calves **2**:207
friendship alliances
2:206
hunting by **2**:205
neck **2**:195–196
circulatory and
respiratory systems
2:189, 202–203
color **2**:193

common **2**:*188*, 190,
193
vertebrae **2**:195
digestive and excretory
systems **2**:189,
204–205
DNA collection from
2:206
echolocation **2**:193,
200, 201
external anatomy
2:189, 191–193
eyes **2**:193
flippers **2**:197
hunting by **2**:200, 205
intelligence **2**:201
long-beaked common
2:*188*, 193
muscular system **2**:189,
197–198
nervous system **2**:189,
199–201
oceanic **2**:*188*, 190
play **2**:*207*
reproductive system
2:189, 193,
206–207
respiratory system
2:194
Risso's **2**:204
river **2**:*188*, 190, 193,
205
skull **2**:*195*
senses **2**:200–201
short-beaked common
2:*188, 191, 194*,
198
circulatory system
2:*202*
digestive and
excretory systems
2:*204*
musculature **2**:*197*
nervous system **2**:*199*
reproductive systems
2:*206*
skull **2**:*196*

skeletal system **2**:189,
194–196
skin **2**:191, *192*
sounds **2**:200
spotted **2**:*200*; **8**:*1031*
swimming **2**:192, 197,
198
tail **2**:197
taste **2**:200
teeth **2**:*189, 195*, 204
douroucouli **6**:*729*
dugong **2**:248; **5**:*700*,
701, 702–703
birth **5**:717
food and digestive
system **5**:713, *714*
mating **5**:716
senses **5**:710
skeleton **2**:250; **5**:*706*
skull **5**:*705*
snout **5**:*704*
tail **5**:*703*
duiker **10**:*1360*, 1361

E
ear ossicles **4**:555
echidna (spiny anteater)
7:*882*, 888
brain **7**:*890*
long-beaked **7**:*882*
long snout of **7**:*886*,
888, 889
penis **7**:*895*
pouch **7**:*897*
remnants of spurs
7:887
short-beaked **7**:*882*,
883, *886*
skeletal system
7:888–889
edentates (xenarthrans)
3:313, 321
eland **10**:1361, *1364*,
1376
elephant **2**:174, **244–263**
African **2**:*244*, 245,
246

body shape **2**:246
feet **2**:249
skull **2**:251
trunk **2**:*246*, 247
tusks **2**:247
African forest **2**:*244*
communication
2:257
size **2**:246
African savanna **2**:*244*
circulatory system
and blood **2**:258
ears **2**:248
feet **2**:*252*
muscular system
2:*254*
in musth **2**:263
nervous system
2:*256–257*
reproductive system
2:*262*
size and weight **2**:246
skull **9**:*1171*
teeth **2**:*261*
trunk and head
2:*247*, 255, *257*
tusks **2**:247, *251*
vertebrae **2**:251
ancient **2**:247
Asian **2**:*244*, 245
body shape **2**:246
brain **2**:256–257
feet **2**:249
food intake **2**:260
skull **2**:251; **9**:*1171*
teeth **2**:*261*
tool use **2**:257
trunk and head
2:*247*
tusks **2**:*251*
vertebrae **2**:251
weight **2**:246
circulatory and
respiratory systems
2:245, 258–259
communication **2**:*253*,
257, 262, 263

elephant *(continued)*
  digestive system 2:245,
    260–261
  dung 2:261
  ears 2:248
    flapping 2:*254*, 255
  energy efficiency 2:259
  external anatomy
    2:245, *246–249*
  extinct 4:511
  feet 2:*252*, 253
  heart rate 2:259
  lymphatic system 2:259
  meeting mates 2:262
  moving 2:252–253
  muscular system 2:245,
    254–255
  in musth 2:262, 263
  nervous system 2:245,
    256–257
  reproductive system
    2:245, 262–263
  sense organs 2:257
  size 2:246, 249
  skeletal system 2:245,
    250–253
  skin 2:248
  teeth 2:*174*, 260–261
  tool use 2:257
  trunk 2:*246*, 247,
    249
    muscles 2:255
  tusks 2:*174*, 247–248,
    251
  water intake 2:260
embryo 7:*1005*
equids 10:1419
  brain 10:*1430–1431*
ermine (stoat; short-
    tailed weasel)
    10:1326–1327,
    1328, *1329*, *1331*,
    1342
  reproduction 10:1343
  size 10:1340
  skull 10:*1333*
  strength 10:1334

estrogen cycle 2:*269*
exocrine glands 2:*274*

F
ferret 10:*1324*, 1328
  black-footed 10:1328
  European 10:*1342*
    skull 10:*1333*
fisher 10:1330, 1342
flying fox 3:*293*
  aerodynamics 3:297
  tail 3:*296*
flying lemur 3:*297*
fox 6:*844*; 10:*1382*,
    1387
  arctic 10:1391
  red 10:1387
fruit bat **3:292–311**
  circulatory and
    respiratory systems
    3:293, *306–307*
  digestive and excretory
    systems 3:293,
    308–309
  Egyptian 3:*305*, 306,
    309
  epauletted 3:*292*, 311
  external anatomy
    3:293, *294–297*
  gray-headed 3:*293*,
    294, 295, 298, 306,
    308, 311
    flight 3:297, *301*
    nerves inside head of
    3:*303*
  hammer-headed 3:*292*,
    311
  Indian short-nosed
    3:*309*
  leaf-nosed 3:*294*
  muscular system 3:293,
    301–302
  nervous system 3:293,
    303–305
  New World (microbats)
    3:*292–293*, 294
    tail 3:*296*

Old World (megabats)
    3:*292*, 293, *294*
  fruit-eating 3:*292*,
    293
  nectar-eating 3:*292*,
    293
  pygmy 3:293
  reproductive system
    3:293, *310–311*
  rousette 3:*292*, 304
  skeletal system 3:293,
    298–300
  tube-nosed 3:*292*

G
gazelle 10:*1360*
  forelimbs 3:*421*
gelada 6:725
giant anteater **3:312–331**
  circulatory and
    respiratory systems
    3:313, *329*
  digestive and excretory
    systems 3:313,
    325–326
  external anatomy
    3:313, *314–317*
  muscular system 3:313,
    322–324; 6:*763*
  nervous system 3:313,
    327–328
  reproductive system
    3:313, *330–331*
  skeletal system 3:313,
    318–321
gibbon 1:*88*, 89, 95, 96;
    4:*550*
giraffe **3:356–373**; 7:*964*;
    10:*1360*
  circulatory and
    respiratory systems
    3:357, *368–369*
  digestive and excretory
    systems 3:357,
    370–371; 10:*1377*
  external anatomy
    3:357, *358–361*

Masai 3:*359*
  muscular system 3:357,
    364–365
  nervous system 3:357,
    366–367
  Nubian 3:*359*
  reproductive system
    3:357, *372–373*
  reticulated 3:*358*, 359
  skeletal system 3:357,
    362–363; 9:*1173*
  teeth 2:*174*
giraffids 3:*356*, 357, 361
glires 4:*476*, 477
glyptodonts 3:320
goat 3:*356*, 357; 7:*964*;
    10:*1360*
gopher, pocket 7:*946*
gorilla 1:*88*, 89, 95, 103;
  teeth 4:569
gray whale **3:374–395**
  circulatory and
    respiratory systems
    3:375, *389–391*
  digestive and excretory
    systems 3:375,
    392–393
  external anatomy
    3:375, *376–379*
  eyes 3:387
  muscular system 3:375,
    383–384
  nervous system 3:375,
    385–388
  reproductive system
    3:375, *394–395*
  skeletal system 3:375,
    380–382
grison 10:*1324*, 1327,
    1329
grizzly bear **3:410–429**
  circulatory and
    respiratory systems
    3:411, *424–425*
  digestive and excretory
    systems 3:411,
    426–427

external anatomy
  3:411, *412–415*
muscular system 3:411,
  420–421
nervous system 3:411,
  422–423
reproductive system
  3:411, *428–429*
skeletal system 3:411,
  416–419
guenon 6:*724*, 725
  hands and feet 6:*728*

H
hamster 7:*946*
  tail 7:*955*
hare **4:476–493**
  arctic 4:*479*, 481, 489
  brown 4:*477*,
    *478–479, 480,
    482, 483, 484,
    489, 490*
    gestation period
      4:*492*
    young 4:*493*
  circulatory and
    respiratory systems
    4:*477, 487–488*
  digestive and excretory
    systems 4:477,
    489–491
  external anatomy
    4:*477, 478–481*
  mountain 4:481
  muscular system 4:477,
    484–485
  nervous system 4:477,
    486
  reproductive system
    4:*477, 492–493*
  skeletal system 4:477,
    482–483
  snowshoe 4:*479,
    481*
hartebeest, hooves
  10:*1366*, 1367
heart 1:112; **4**:*487*

hippopotamus 3:*356,
  357*; **4:506–525**;
  7:*964*; **10**:*1360*
  circulatory and
    respiratory systems
    4:*507, 520–521*
  common 4:*506*, 508,
    *509, 510, 512, 514,
    516*, 517, *518*, 519,
    520, 522, 523,
    524–525
  digestive and excretory
    systems 4:507,
    522–523
  external anatomy
    4:507, *508–511*
  muscular system 4:507,
    516–517
  nervous system 4:507,
    518–519
  pygmy 4:*506, 509,
    510, 523*
  reproductive system
    4:507, *524–525*
  skeletal system 4:507,
    512–515
hog 4:*506*
hominids 4:*550*, 551
  extinct, diet 4:*557*
  foramen magnum
    4:556
hominines 4:*550*, 551
horse 8:*1032*; **10**:1419
  Clever Hans **10**:1431
  domestication
    10:1423
  geophagia **10**:1434
  horse-ass hybrids
    10:1437
  hybrids **10**:1437
  jaw **10**:*1425*
  Przewalski's **10**:*1418*,
    1423
  sounds **10**:1433
  teeth **10**:1435
  ventilation at speed
    8:1031

human 1:*88*, 89;
  4:550–573
  bipedalism 4:552–553,
    556, 558, 561
  circulatory and
    respiratory systems
    4:551, *566–567*
    ventilation 8:*1028*
  digestive and excretory
    systems 4:551,
    568–569
    liver 2:*184*; 4:568
    stomach 2:*177*
  embryo 7:*1004*
  endocrine system
    2:*270*; 4:*570–571*
  external anatomy
    4:551, *552–554*
  eyes 4:*553*; 6:*788*
  heart 1:*113–117*; **4**:*567*
  heart attacks 4:*567*
  life at high altitudes
    4:*566*
  muscular system 4:551,
    559–561
  nervous system 4:551,
    562–565
    brain 6:*794, 795,
      796, 799*
    brain size and
      intelligence 4:*564*
  oral cavity 2:*173*
  reproductive system
    4:551, *572–573*;
    7:*994*–996
  skeletal system 4:551,
    555–558; 9:*1169*
    hand bones 9:*1174*
  tears 2:275
  teeth 4:*557*
  urinary system, male
    2:*185*
  vision 4:*563*
  vocal cords and
    muscles 4:*560*
  whites of the eyes
    4:*553*

hyena 5:**592–605**, *664*;
  7:*928*; **10**:*1324*
  African folklore and
    5:601
  ancestors 5:*595*
  brown 5:*592*, 593, *595*,
    602, 604
  circulatory and
    respiratory systems
    5:*593, 602*
  digestive and excretory
    systems 5:593, *603*
  external anatomy
    5:593, *594–595*
  hunting by 5:*600*
  laughing 5:*602*
  muscular system 5:593,
    598–599
  nervous system 5:593,
    600–601
  reproductive system
    5:593, *604–605*
  scent 5:*595*, 600–*601*
  skeletal system 5:593,
    596–597
    hind limb bones
      9:*1175*
  spotted 5:*592*, 593,
    595, 596, *597, 598,
    601*, 602, 603, *605*
  striped 5:*592*, 593,
    594–595, 596, *597,
    598, 600*, 602, *603*,
    604
  see also aardwolf
hyrax 2:*244*, 245, 249

I
ictithermium 5:*595*
impala **10**:*1367*

J
jackal **10**:*1382*, 1384
  black-backed **10**:*1382*
jackrabbit 4:*476*
  antelope 4:*479*
  black-tailed 4:*479*

jaguar **5**:*664*
  eyes **5**:*668*
jerboa **7**:*946*, 951
  tail **7**:*955*

K

kangaroo **5**:**648–663**
  ancestors **5**:652
  circulatory and
    respiratory systems
    5:649, *658*
  digestive and excretory
    systems **5**:649,
    659–660; **10**:*1377*
  eastern gray **5**:*648*, 649
  external anatomy
    5:649, *650–653*
  gray **5**:*651*, *657*
  movement **5**:*652*, 653,
    656
  muscular system **1**:*56*;
    5:649
  nervous system **5**:649,
    657
  red **5**:*648*, 649, *650*,
    651, *652*, *654*, *655*,
    656, 660, *662–663*
  reproductive system
    5:649, *661–663*;
    7:*996*
  skeletal system **5**:649,
    654–655
    hind limb bones
      9:*1175*
  tree **5**:649, *660*
  western gray **5**:649
kidneys **2**:*186*
klipspringer, hooves
  10:*1366*
koala **5**:*648*
kudu **10**:1361
  horns **10**:*1365*

L

lagomorphs **4**:*476–481*,
  484, 486, 489–491,
  492–493

langur *see* monkey, langur
lemming **7**:*946*
lemur **1**:95; **4**:*550*
  flying **3**:*297*
leopard **5**:*664*, 671
  clouded **5**:*665*; **7**:*932*
leporids **4**:477
lion **5**:603, **664–683**
  African **5**:665, *670*,
    671, *672*, *674*, *676*,
    677, *679*, *681*, *682*
  Asiatic **5**:665
  Barbary **5**:667
  circulatory and
    respiratory systems
    5:665, *677–678*
  cubs **5**:*682*, *683*
  digestive and excretory
    systems **5**:665,
    679–680
  evolution **5**:668
  external anatomy
    5:665, *666–669*
  hunting by **5**:668–669,
    672–673, *674–675*,
    678, *680*
  muscular system **5**:665,
    672–673
  nervous system **5**:665,
    674–676
  prides (groups) of
    5:*680*, 681, 682
  reproductive system
    5:665, *681–683*
  roaring **5**:*671*
  skeletal system **5**:665,
    670–671
llama **7**:*964*; **10**:*1360*
loris **1**:95; **4**:*550*; **6**:738
lynx **7**:*928*

M

macaque **6**:*724*, 725,
  728
  Barbary **6**:725, 727
  brain **6**:737
mammoth **2**:*244*, 245

manatee **2**:248;
  5:**700–717**
  Amazonian **5**:*700*, 703,
    704, 712, 717
  Antillean **5**:701
  buoyancy **5**:711
  circulatory and
    respiratory systems
    5:701, *711–712*
  communication **5**:710
  digestive and excretory
    systems **5**:701,
    713–714
  diving **5**:712
  external anatomy
    5:701, *702–703*
  Florida **5**:701
  muscular system **5**:701,
    708
  nervous system **5**:701,
    709–710
  reproductive system
    5:701, *715–717*
  skeletal system **2**:250;
    5:701, *705–706*;
    9:1168
  West African **5**:*700*,
    701, 702, 703, *704*,
    717
  West Indian **5**:*700*,
    701, *702–703*, 704,
    705, 706, 707, 709,
    713, 715, 717
mandrill **6**:**724–743**
  circulatory and
    respiratory systems
    6:725, *738*
  colors on face and
    buttocks **6**:*725*,
    726, 728, *730*, 738
  digestive and excretory
    systems **6**:725,
    739–741
  external anatomy
    6:725, *726–730*
  muscular system **6**:725,
    734–735

nervous system **6**:725,
  736–737
  reproductive system
    6:725, *742–743*
  skeletal system **6**:725,
    731–733
mangabey **6**:*724*, 725
marmoset
  brain **6**:736
  jaw **6**:741
  skull **6**:*732*
marmot **9**:*1208*
  Alpine **9**:1210
marsupials **1**:*88*; **4**:551;
  5:*648*, 700; **6**:*844*;
  7:885
  cloaca **2**:*185*
  evolution **9**:1193
marten **10**:*1324*, 1325
  American
    10:1329–1330
  beech **10**:*1330*
  pine **10**:1329, 1339,
    1340
  reproduction **10**:1341,
    1343
mastodon **2**:*244*
mazaalai **3**:411
microbats **3**:292–293
  echolocation **3**:305
  tail **3**:*296*
mink **10**:*1324*, 1325,
  1331, 1335
  American
    10:1328–1329,
    1341, 1343
mongoose **5**:*592*, *664*;
  7:*928*; **10**:*1324*
monkey
  catarrhine monkeys
    1:89; **6**:725, 730
  colobus digestive
    system **6**:739
  hands and feet **6**:*728*,
    729
  skull **6**:*732*
  teeth **6**:741

drill **6**:725
dusky titi **6**:*729*
haplorrhine monkeys
    6:725
  langur digestive
    system **6**:739
  jaw **6**:741
  New World **1**:*88*;
    4:*550*; **6**:*724*, 725
    cebids **6**:*729*, 730
    vomeronasal organ
    6:737
  Old World **1**:*88*, 89;
    4:*550*; **6**:*724*, 725
  platyrrhine monkeys
    1:*89*; **6**:730
  proboscis, digestive
    system **6**:739
  red howler **6**:*729*
  spider **6**:*734*, 735
    black spider **6**:*729*
  vervet **6**:*730*
monotremes **4**:551;
    7:*882*, 883, 885
  body temperature
    7:*887*
  brain **7**:891
  cloaca **2**:185; **7**:883,
    894, 895–896
  mammary glands **7**:*896*
  skeletal system
    7:888–889
moose **7**:*967*
mouse **7**:*946*
  coprophagy **7**:961
  differs from rats
    7:*950*–951
  food and feeding **7**:961
  harvest **7**:*955*, 961, 963
  house, skull **7**:*953*
  New World **7**:*946*
  Old World **7**:*946*
  paws **7**:*949*
  wood **7**:961
mule **10**:*1437*
muskrat, tail **7**:*955*
mustelids **10**:1325

N

narwhal **2**:*188*, 190
  jousting **2**:206
Neanderthal people
    4:*550*
nonplacental **7**:882–883
numbat **3**:321
nyala **10**:*1376*

O

ocelot **7**:*928*
okapi **3**:*356*, 357, 359,
    360
orangutan **1**:*88*, 89, 97,
    103; **4**:*550*
orca *see* whale, killer
otter **6**:**844–861**;
    10:*1324*
  ancestry **6**:847
  Asian short-clawed
    6:*849*, 852, 858
  circulatory and
    respiratory systems
    6:845, *856–857*
  clawless **6**:849, 851,
    852, 855, 858, 861
  digestive and excretory
    systems **6**:845,
    858–859
  Eurasian river **6**:*844*,
    847, *848*, *859*,
    860–861
  external anatomy
    6:845, *846–849*
  eyes **6**:848–849, 853,
    855
  fur **6**:847–848, 849,
    850
  giant **6**:*844*, 845, 847,
    848
  Lontra-type **6**:*844*, 845
  Lutra-type **6**:*844*, 845
  muscular system **6**:845,
    853
  nervous system **6**:845,
    854–855

North American river
    6:*844*, *845*, *848*,
    861
  number of species
    6:845
  reproductive system
    6:845, *860–861*
  scent **6**:854, 860
  sea **6**:*844*, 845, *846*,
    847, *849*, *851–861*
  sizes **6**:*849*
  skeletal system **6**:845,
    850–852
  sociable or solitary
    6:861
  streamlining **6**:848
ox, Vu Quang **3**:360

P

panda, giant **3**:*410*, 411,
    419
pangolin **3**:315; **9**:1185
  Cape **9**:*1181*
panther, Florida
    7:944
paranthropines **4**:*550*
peccary **4**:507
pig **3**:*356*; **4**:507; **7**:*964*;
    10:*13600*
pika **4**:476, 477, 479,
    480
  skull **4**:*483*
  steppe **4**:*479*
pinnipeds **8**:1111
  origins **8**:*1113*
placental **1**:*88*, 89; **2**:189,
    245; **3**:*356*, *410*;
    4:477, *506*, 550;
    5:*648*, 665, *700*;
    6:*724*, 844; **7**:*899*,
    966; **8**:*1032*, 1033;
    10:1325, 1361,
    1418
platypus **7**:**882–897**
  ancestors **7**:886
  bill **7**:*886*–887, 889,
    892, 894

circulatory and
    respiratory systems
    7:883, *893*
  digestive and excretory
    systems **7**:883, *894*
  diving by **7**:893
  electroreceptors **6**:*790*
  external anatomy
    7:*883*, *884–887*
  first discovered **7**:885
  muscular system **7**:883,
    890
  nervous system **7**:883,
    891–892
  reproductive system
    7:883, *895–897*
  skeletal system **7**:883,
    888–889
  venomous spurs **7**:*887*,
    890
polecat **10**:*1324*, 1325,
    1328
  African (zorilla)
    10:1330–1331,
    1336
  European **10**:1335,
    1336, *1343*
    skull **10**:*1333*
  feet **10**:*1329*
  marbled **10**:1330,
    1336, 1341
  reproduction **10**:1341
  thin **10**:1330
porcupine **7**:**898–915**
  African **7**:*902*, 905,
    907, 914–915
  Brazilian **7**:906
  brush-tailed **7**:900, 902
  circulatory and
    respiratory systems
    7:899, *910–911*
  crested **7**:900, 904–*905*
  digestive and excretory
    systems **7**:899,
    912–913
  external anatomy
    7:899, *900–903*

porcupine *(continued)*
  hairy dwarf **7**:902, 912
  Indonesian **7**:902
  long-tailed **7**:900, 902
  muscular system **7**:899,
    906–907
  nervous system **7**:899,
    908–909
  New World **7**:*898*,
    901, *902–903*
  North American
    7:*900–906, 908,*
    910, *912–915*
    craving for salt **7**:*913*
  Old World **7**:*898, 901,*
    912
  pheromones **7**:908, 914
  prehensile-tailed **7**:*899,*
    912
  reproductive system
    7:899, *914–915*
  skeletal system **7**:899,
    904–905
  South American tree
    7:*903*
  spines (quills) and
    rattles **7**:*899,*
    900–901, 907
  stump-tailed **7**:*903*
  three crested **7**:*901*
porpoise **2**:*188*, 190
  harbor, skull **2**:*195*
  teeth **2**:*195*, 204
possum **5**:*648*, 649
potoroo **5**:649
prairie dog **8**:1026;
  9:*1208*
primates **1**:*88*, 89; **4**:*550*,
  551; **6**:*724–725*
  ancestors **6**:727
  diets **1**:101
  heart **6**:738
  limbs **6**:733
  locomotion **6**:734–735
  manual dexterity
    6:733

reproductive biology
  6:742
senses **6**:737
  color vision and
    frugivory **6**:740
  skulls **6**:731
  stereoscopic vision
    6:*736*
  varied lifestyles **6**:729
proboscideans **2**:247
pronghorn **3**:*356*; **7**:*964*;
  10:*1360, 1371*
prosimians **6**:737
pudu **7**:966
puma **7**:**928–945**
  circulatory and
    respiratory systems
    7:929, *940–941*
  digestive and excretory
    systems **7**:929,
    942–943
  external anatomy
    7:929, *930–933*
  killing prey **7**:943
  muscular system **7**:929,
    936–937
  nervous system **7**:929,
    938–939
  power and agility
    7:936, *937*
  reproductive system
    7:929, *944–945*
  skeletal system **7**:929,
    934–935
  whiskers **7**:*933*

Q
quokka **5**:649, 653, 656

R
rabbit **4**:*476*
  Annam striped **4**:*476*
  bristly **4**:*476*
  bushman **4**:*476*
  Central African **4**:*476*
  European **4**:*476, 479*

pygmy **4**:*476*
  red **4**:*476*
  Ryuku **4**:*476*
  short-eared **4**:*476*
  skull **4**:*483*
  Sumatran short-eared
    4:*476*
  volcano **4**:*476*
  young **4**:*493*
raccoon **3**:*410*; **5**:*592,*
  664; **7**:*928*;
  10:*1324*
rat **7**:**946–963**
  ancestors **7**:949
  black **7**:950
  brown **7**:*947–963*
  circulatory and
    respiratory systems
    7:947, *958–959*
  communication **7**:956
  dentition **7**:*950*, 953
  digestive and excretory
    systems **7**:947,
    960–961
  external anatomy
    7:947, *948–951*
  kangaroo **7**:*951*
  muscular system **7**:947,
    954–955
  nervous system **7**:947,
    956–957
  New World **7**:*946*
  Old World **7**:*946*
  poisons and **7**:960
  pouched **7**:950, *961*
  reproductive system
    7:947, *962–963*
  skeletal system **7**:947,
    952–953
  used for
    experimentation
    7:957
rat-kangaroo **5**:649
red deer (wapiti)
  7:**964–983**
  bellowing **7**:*978*

circulatory and
  respiratory systems
  7:965, *977–978*
digestive and excretory
  systems **7**:965,
  979–980
external anatomy
  7:965, *967–969*
metabolic slow-down
  7:978
muscular system **7**:965,
  973–974
nervous system **7**:965,
  975–976
reproductive system
  7:965, *981–983*
skeletal system **7**:965,
  970–972
subspecies **7**:968
reedbuck **10**:*1366*
reindeer *see* caribou
reproductive systems
  994–996
respiratory systems
  gas exchange **8**:*1012*,
    1019–1020
  ventilation **8**:1028
rhinoceros **8**:**1032–1047**;
  10:*1418*, 1419
  black **8**:*1032*, 1033,
    1035–1036, *1037*,
    1039, 1041, 1047
  circulatory and
    respiratory systems
    8:1033, *1043*
  digestive and excretory
    systems **8**:1033,
    1044–1045
  external anatomy
    8:1033, *1034–1037*
  horns **8**:*1035–1036*
  Indian **8**:*1032*, 1035,
    1036, 1037, 1045,
    1047
  Javan **8**:*1032*, 1033,
    1034, 1035, *1037*

muscular system
8:1033, *1040–1041*
nervous system 8:1033, *1042*
prehistoric 8:*1036*
reproductive system
8:1033, *1046–1047*
skeletal system 8:1033, *1038–1039*; 10:*1426*
hind limb bones 9:*1175*
subspecies 8:1037
Sumatran 8:*1032*, 1033, 1034, 1035, *1036, 1037*, 1046
white 8:*1032–1047*
rodents 4:*476*, 477; 7:*898, 946*, 947; 9:*1208*, 1209
ancestors 7:949
cavylike 7:*898*, 899, 907, *946*, 947, 952
mouselike 7:*898*, 907, *946*; 9:*1208*
squirrel-like 7:*898*, 907, *946*; 9:*1208*
tail 7:955
teeth 4:482; 7:905
rorquals 3:375
calves 3:395
feeding 3:392–393
ruminants 3:*356*; 4:*506*, 522; 7:*964*, 979–980; 10:*1360*
sleep 10:1373
stomach 2:177; 7:979

S
sable 10:1329
horns 10:*1365*
saiga 10:*1360*
sea cat 6:*848*
sea cow 2:*244*, 245
blubber 2:192
Steller's 5:*700*

see also dugong; manatee
seal 3:*410*; 6:*844*; 8:**1110–1125**; 10:*1382*
blubber 2:192
circulatory and respiratory systems 8:1111, *1120–1121*
common 8:*1115*
crabeater 8:1122
digestive and excretory systems 8:1111, *1122–1123*
diving by 8:*1013*, 1117, 1121
eared 8:1111, 1113, 1115, 1116
see also sea lion
elephant 8:*1013*, 1113, *1114, 1115*, 1118, 1121, 1122, 1123, 1125
external anatomy 8:*1111, 1112–1115*
eyes and vision 8:1114, 1118, 1119
fasting by 8:1123
fur 8:*1110*, 1111, *1115*, 1116
gray 8:*1123, 1125*
harbor 8:1111, *1112*, 1113, *1114*–1115, *1116*, 1117, *1120*, *1122, 1124*
harp 8:*1114*
Hawaiian monk 8:1112
hearing underwater 8:1119
hooded 8:*1121*
leopard 8:*1115*
muscular system 8:1111, *1117*
nervous system 8:1111, *1118–1119*
northern 8:*1110*, 1111

reproductive system 8:1111, *1124–1125*
ringed 8:1112, 1113, 1116, 1117, 1122
skeletal system 8:1111, *1116*
sounds 8:*1121*
southern 8:*1110*, 1111
tasting saltiness of seawater 8:1119
true (earless) 8:*1110*, 1113, 1115, 1116, *1117*
Weddell 8:1112, *1113*, 1117, 1120, 1121
whiskers 8:1114, *1119*
white-coated true 8:1111
sea lion 3:*410*; 8:*1110*, 1111; 10:*1382*
blubber 2:192
leg muscles 8:*1117*
sheep 3:357
shrew, tree *see* tree shrew
sirenians 5:700–701
see also dugong; manatee
sitatunga, hooves 10:*1366*
skeletal system 9:1160, 1170
skunk, odor 2:275
sloth 3:*312*, 313; 9:**1176–1193**
ancestors 9:*1182*
brown-throated three-toed 9:*1176*, 1180, 1183, *1189, 1191*
circulatory and respiratory systems 9:1177, *1188–1189*
claws 3:316
digestive and excretory systems 9:1177, *1190–1191*
external anatomy 9:1177, *1178–1183*

fur
cyanobacteria in 9:*1178*, 1182, 1183
moths in 9:1182
giant 3:320
Hoffman's two-toed 9:*1176*, 1179–1180, *1183*, 1193
maned three-toed 9:*1176*, 1180, 1182–1183, 1190
muscular system 9:1177, *1186*
nervous system 9:1177, *1187*
pale-throated three-toed 9:*1176, 1177, 1178*, 1180, 1183, *1184, 1186, 1188, 1190*
reproductive system 9:1177, *1192–1193*
skeletal system 9:1177, *1184–1185*
southern/Linne's two-toed 9:*1176*, 1179–1180, *1187*
three-toed 3:*312*, 313; 9:*1176*, 1177, *1179, 1180*, 1181, *1185*
two-toed 3:*312*, 313; 9:*1176*, 1177, *1179, 1180, 1185*
squirrel 9:**1208–1223**
Abert's 9:1212
African pygmy 9:1210
arctic ground 9:1211, *1215*
Arizona gray 9:*1221*
black giant 9:1210, *1211*
circulatory and respiratory systems 9:1209, *1219*

squirrel (continued)
   digestive and excretory
      systems **9**:1209,
      1220–1221
   Douglas (chickaree)
      8:*1148*
   eastern gray **9**:1210,
      1220, 1223
   Eurasian red
      9:*1210–1216,*
      1218, 1219, 1220,
      1222, 1223
   evolution and
      relationships **9**:1215
   external anatomy
      9:1209, *1210–1213*
   feet **7**:950; **9**:1211
   feeding on fungus
      9:1221
   flying **9**:*1208, 1213*
      tail **7**:*955*; **9**:*1217*
   giant flying **9**:*1212,*
      1213
   gray **9**:*1212,* 1221,
      1223
   groove-toothed **9**:1212
   ground **9**:*1208,*
      1211–*1212,* 1220
   muscular system
      9:1209, *1216–1217*
   nervous system **9**:1209,
      1218
   northern flying **9**:1223
   pygmy tree **9**:1217
   red **9**:*1209,* 1221
   reproductive system
      9:1209,
      1222–1223
   rock **9**:1223
   scaly-tailed **9**:*1208,*
      1209, 1213
   shrew-faced ground
      9:1212
   skeletal system **9**:1209,
      1214–1215
   southern flying **9**:1213,
      1217

   tail of **9**:1212, 1217
   tree **9**:*1208,*
      1211–1212, 1220
stoat *see* ermine

T

tamandua **3**:*312,* 313
   anal secretions **3**:327
   claws **3**:316
   collared **3**:316
   northern **3**:316, *317*
   skull **3**:*319*
   southern **3**:316, *317,*
      328
   tail **3**:317
   teeth **3**:321
tapir **8**:*1032*; **10**:*1418,*
   1419
   skull **10**:*1426*
tarsier **4**:*550*; **6**:731
tayra **10**:*1324,* 1330,
   1339
thrombosis **1**:*117*
tiger **5**:*664, 668*
tree shrew **6**:727

U

ungulates **7**:966;
   10:*1361,* 1366
   even-toed (artiodactyls)
      3:*356,* 357, 377;
      4:*506*–507, *564*;
      7:966; **8**:*1032*;
      10:*1360,* 1361,
      1366, *1418*
   piglike **4**:507
   feet **10**:1369
   mammary glands **2**:*275*
   odd-toed
      (perissodactyls)
      7:966; **8**:*1033*;
      10:*1360,* 1361,
      1366, *1418*–1419,
      1421
   as relatives of whales
      2:192

V

vole **7**:*946*

W

wallaby **5**:649, 650
   black-footed rock
      5:*653*
   red-necked **5**:*648*
   rock **5**:649
wallaroo **5**:*648,* 649, *651*
walrus **3**:*410*; **8**:*1110,*
   1111; **10**:*1382*
   blubber **2**:192
   tusks **8**:1125
wapiti *see* red deer
weasel **3**:*410*; **5**:*592,*
   664; **6**:845; **7**:*928*;
   10:**1324–1343**,
   1325, *1382*
   African striped
      10:1330
   circulatory and
      respiratory systems
      10:1325,
      1337–1338
   digestive and excretory
      systems **10**:1325,
      1339–1340
   external anatomy
      10:1325,
      1326–1331
   least **6**:845;
      10:*1324–1327,*
      1329, 1332–1333,
      1335, 1337,
      1338–1343
   long-tailed **10**:1326,
      1327, 1342
   muscular system
      10:1325, *1334*
   nervous system
      10:1325,
      1335–1336
   northern/snow
      10:1327–1328,
      1340

   parasitic worms and
      10:1332
   reproductive system
      10:1325,
      1341–1343
   short-tailed *see* ermine
   size **10**:1340
   skeletal system
      10:1325,
      1332–1333
   smelly defense **10**:1336
whale **4**:*506*
   ancient forerunners of
      2:192
   baleen **2**:*188,* 189;
      3:*374–395*
      relatives **3**:377
   beaked **2**:*188,* 190
   Blainville's beaked
      3:382
   blubber **2**:191, 192
   blue **3**:*374,* 377, *379*
      calves **3**:395
      eyes **3**:*387*
      feeding **3**:*392*
      stomach **3**:393
   bones **2**:194
   bowhead **3**:378–379,
      388, 389
   Bryde's **3**:*374*
   deep-diving **2**:202, 203
   DNA collection from
      2:206
   fin **3**:*374,* 378
      ears **3**:*388*
   gray *see* gray whale
   heat exchanger **2**:202
   humpback **3**:*374, 377,*
      378
      barnacles on **3**:378
      hunting by **2**:200
      singing **3**:387, 388
      skull **3**:*381*
   hunting by **2**:200–201,
      205
   killer (orca); **2**:190
      3:*377*

communication 2:201
dorsal fins 2:190
hunting by 2:205
skull 2:195
teeth 2:196, 204
minke 3:388
neck 2:195–196
pilot 2:190
right 3:374, 375
barnacles on 3:378
Northern right 3:374
pygmy right 3:374, 395
skull 3:381
Southern right 3:374, 395
rorqual, filter feeding 2:175
senses 2:200–201
skeletal system, thoracic vertebrae 9:1172
skin 2:191
sperm whales 2:188, 189
ambergris 2:205
deep-diving 2:203
DNA collection from 2:206
hunting by 2:200
muscles 2:197
sexual maturity 2:207
teeth 2:204
swimming 2:198; 3:376, 379, 384
toothed 2:188, 189; 3:374, 375
brain 2:199–200
breathing 2:192
communication 2:201
digestion by 2:204
prey 2:204
reproduction 2:206–207

senses 2:193, 200–201
skull 2:195
see also beluga; blackfish; narwhal
wildcat 7:928, 945
wildebeest 10:1360–1381
black 10:1360
carriers of disease 10:1374
circulatory and respiratory systems 10:1361, 1374
common/blue 10:1360, 1361, 1362, 1368, 1370, 1372, 1374, 1378
digestive and excretory systems 10:1361, 1375–1377
external anatomy 10:1361, 1362–1367
migration 10:1364
muscular system 10:1361, 1370–1371
nervous system 10:1361, 1372–1373
reproductive system 10:1361, 1378–1381
skeletal system 10:1361, 1368–1369
wolf 6:844; 10:1382–1403
circulatory and respiratory systems 10:1383, 1397
communication 10:1388, 1389, 1401–1402
cubs 10:1402, 1403

digestive and excretory systems 10:1383, 1398–1399
Ethiopian 10:1382, 1403
external anatomy 10:1383, 1385–1389
gray 10:1382–1387, 1390–1402
maned 10:1382, 1385, 1391
muscular system 10:1383, 1392–1393
nervous system 10:1383, 1394–1396
packs 10:1400–1402
red 10:1382, 1384, 1388
regurgitation by 10:1399, 1403
reproductive system 10:1383, 1400–1403
skeletal system 10:1383, 1390–1391
teeth 2:174
timber 10:1385
tracks 10:1389
vocalizations 10:1397
wolverine 3:419; 6:845; 10:1324, 1325, 1330, 1331, 1335
wombat 5:648

X

xenarthrans (edentates) 3:313, 321

Z

zebra 8:1032; 10:1418–1437
alertness 10:1429, 1431
ancestors 10:1421

avoiding foes 10:1427
circulatory and respiratory systems 10:1419, 1432–1433
digestive and excretory systems 10:1419, 1434–1435
external anatomy 10:1419, 1420–1423
foals 10:1436, 1437
Grevy's 10:1422, 1433, 1435
mountain 10:1419, 1422, 1433
muscular system 10:1419, 1427–1428
nervous system 10:1419, 1429–1431
plains 10:1419, 1420, 1422, 1425–1429, 1432, 1433, 1434, 1435, 1436
reproductive system 10:1419, 1436–1437
skeletal system 10:1419, 1424–1426
sounds 10:1433
stripes 10:1420–1421, 1422, 1423
zorilla (African polecat) 10:1330–1331, 1336

Plants, fungi, and algae

A

algae
  brown **8**:*1126*, **8**:1127,
    8:1129, **8**:*1130*,
    8:1131, **8**:1133
  green **2**:164; **8**:*1126*,
    8:1127, **8**:1131
  life cycles
    8:1134–**8**:1135
  red **8**:*1126*, **8**:1127,
    8:1131, **8**:1133
  reproduction **8**:*1134*,
    8:1135–**8**:1136
  see also seaweed
alternation of
    generations **7**:*993*
angiosperms **1**:*36*, **1**:37;
    6:*744*, **6**:*824*;
    7:*916*, **7**:917;
    8:*1048*, **8**:*1138*;
    10:*1300*
  reproductive structures
    7:1000–**7**:*1003*
apple tree **1**:**36**–**1**:**45**
  anatomy
    external **1**:37,
      1:*38*–**1**:*39*
    internal **1**:37,
      1:*40*–**1**:*42*; **9**:*1161*
  reproductive system
    1:37, **1**:*43*–**1**:*45*;
    7:*1001*
apricot **1**:44

B

bean **1**:*36*
belladonna **7**:*916*,
    7:*925*
bird's nest fungi **6**:771,
    6:*778*
blackberry (bramble)
    1:*38*, **1**:*39*
bladderwort **10**:1305

bracken **2**:*276*, **2**:*279*,
    2:*282*
bracket fungi **6**:771,
    6:*773*, **6**:*775*, **6**:*778*
butterwort **10**:1303

C

cactus **8**:*1048*
  claret cup **8**:*1059*
  and drought
    8:1056–**8**:*1057*
  external anatomy
    8:1049,
    8:*1050*–**8**:*1054*
  golden barrel **8**:*1050*
  internal anatomy
    8:1049,
    8:*1055*–**8**:*1057*
  reproductive system
    8:*1058*–**8**:*1059*
  saguaro *see* saguaro
    cactus
  seeds **8**:1058, **8**:1059
  surface area and desert
    survival **8**:1054
carnivorous plants
    10:1300–**10**:1301,
    10:1305carrot **1**:*36*
cells **1**:67; **6**:*829*
cherry **1**:37, **1**:38, **1**:43,
    1:44
chitrids **6**:770–**6**:771
club moss **2**:276, **2**:284
conifer **8**:*1138*–**8**:1139
  external anatomy
    8:*1140*–**8**:*1143*
  internal anatomy
    8:*1144*–**8**:*1145*
  reproductive system
    8:*1146*–**8**:*1149*
cordgrass **6**:*744*, **6**:*745*
  common **6**:*744*
  small **6**:*744*
  smooth *see* marsh grass
corn (maize) **6**:*749*,
    6:*753*
cotton **6**:752

crassulacean acid
    metabolism (CAM)
    6:830; **8**:*1057*
crust fungus **6**:*779*
cup fungus **6**:*774*, **6**:775
cycad **8**:*1138*
cypress **8**:*1138*
  California **8**:1142
  Montezuma **8**:1139
  swamp **8**:1143

D

dandelion, seeds **7**:*988*
dead man's fingers
    8:*1133*
death cap **6**:771
destroying angel **6**:771
dicots **1**:*36*, **1**:37; **6**:*744*,
    6:*824*; **7**:*916*,
    7:917; **8**:*1048*;
    10:*1300*–**10**:1301

E

earth star **6**:771
euphorbia **8**:1052,
    8:*1054*

F

fern **2**:**276**–**2**:**285**
  adder's-tongue **2**:*276*,
    2:*280*
  energy store **2**:282
  eusporangiate **2**:*276*
  external anatomy
    2:277, **2**:*278*–**2**:*280*,
    2:*278*–**2**:*280*
  families **2**:277
  filmy **2**:*276*
  internal anatomy
    2:277, **2**:*281*–**2**:*282*,
    2:*281*–**2**:*282*
  lady **2**:*276*, **2**:*277*,
    2:*278*
  leaves (fronds)
    2:*278*–**2**:*279*
  growth **2**:282
  leptosporangiate **2**:*276*

reproductive system
    2:**277**, **2**:**283**–**2**:**285**,
    2:*283*–**2**:*285*
  rhizomes **2**:*279*–**2**:*280*,
    2:282, **2**:285
  stag's-horn **2**:*276*,
    2:280
  tree **2**:279–**2**:280
  water ferns **2**:280
fir **8**:*1138*
firethorn **1**:*36*
flask fungi **6**:771
flowering plants **7**:917
fly agaric **6**:*771*, **6**:*772*,
    6:775
fungi **6**:*770*
  catching nematodes
    6:*774*
  hormones **2**:269

G

ginkgo **8**:*1138*
gnetophyte **8**:*1138*
grapevine **1**:*36*
grass **6**:*744*–**6**:*745*
  marsh *see* marsh grass
  pampas **6**:751
growth hormones **2**:266
gymnosperms **1**:*36*, **1**:37;
    6:*744*, **6**:*824*;
    7:*916*, **7**:1000;
    8:*1048*, **8**:*1138*;
    10:*1300*
  reproduction in
    7:1003

H

hawthorn **1**:*36*, **1**:37,
    1:38
horsetail **2**:*276*, **2**:*279*,
    2:284
  hairy **2**:279

J

jelly fungi **6**:771
juniper, Utah **8**:1144

K

kauri **8**:1142
kelp **8**:*1129*, **8**:*1131*
  bullwhip **8**:1130,
    8:1131
  feather boa **8**:*1129*,
    8:1130
  giant **8**:*1127*, **8**:*1132*
  sea urchins and **6**:859
  winged **8**:1130

L

larch **8**:1140
  bark **8**:*1141*
  cones **8**:*1143*
  leaves **8**:*1142*
laver, purple **8**:*1126*,
  8:*1127*, **8**:*1129*
lichen **6**:773
lily **6**:*744*
  cobra **10**:1301, **10**:1305
lower vascular plants
  1:*36*; **2**:276–**2**:*277*;
  6:*744*, **6**:*824*;
  7:*916*; **8**:*1048*;
  10:*1300*

M

maize (corn) **6**:*749*,
  6:*753*
marsh grass **6**:**744**–**6**:**753**
  external anatomy
    6:745, **6**:*746*–**6**:*748*
  internal anatomy
    6:745, **6**:*749*–**6**:*750*
  reproduction **6**:745,
    6:*751*–**6**:*753*;
    7:*1001*
medlar **1**:*36*
mildew, powdery **6**:771
monkey puzzle tree
  8:1139, **8**:*1140*,
  8:1142
monocots **1**:*36*, **1**:37;
  6:*824*–**6**:*825*;
  7:*916*, **7**:917;
  8:*1048*; **10**:*1300*

roots **6**:826–**6**:827
seeds **6**:*753*
morel **6**:*774*
moss **2**:285
mushroom **6**:**770**–**6**:**779**
  external anatomy
    6:771, **6**:*772*–**6**:*775*
  internal anatomy
    6:771, **6**:*776*
  "magical" growth of
    6:776
  reproduction **6**:771,
    6:*777*–**6**:*779*
mycorrhizae **6**:773,
  6:831; **8**:1143

N

nonvascular plants **1**:*36*;
  6:*744*, **6**:*824*;
  8:*1048*; **10**:*1300*

O

oomycota **6**:770
orchid **6**:*744*,
  6:**824**–**6**:**833**
  bee **6**:832
  bog **6**:832
  epiphytic **6**:825,
    6:826, **6**:827,
    6:830, **6**:*831*
  external anatomy
    6:825, **6**:*826*–**6**:*828*
  grass pink **6**:825,
    6:*827*, **6**:*832*,
    6:*833*
  internal anatomy
    6:825, **6**:*829*–**6**:*831*
  reproduction **6**:825,
    6:*832*–**6**:*833*
  slipper **6**:832
  underground **6**:827,
    6:*828*

P

peach **1**:37, **1**:*44*
pear **1**:37
  European **1**:*36*

pine **8**:*1138*
  bristlecone **8**:1139,
    8:1145
  cones **7**:*1000*, **7**:*1003*;
    8:*1146*, **8**:*1147*
  Monterey **8**:1148
  Wollemi **8**:1139
pitcher plant **10**:1301,
  10:*1303*, **10**:1307,
  10:1311
  American **10**:*1300*,
    10:1305
  northern **10**:*1302*
  swamp **10**:1303,
    10:1304
  Western Australian
    10:*1300*, **10**:*1304*
plum **1**:37, **1**:44
poisons **2**:187
  curare **6**:761
  in the potato family
    7:*925*
polyploid plants **7**:927
pome **1**:*36*, **1**:44
potato **7**:**916**–**7**:**927**
  eyes **7**:*920*, **7**:*921*,
    7:*923*
  flowers **7**:*921*, **7**:*926*,
    7:*927*
  fruit **7**:*922*, **7**:*926*,
    7:927
  reproduction **7**:917
    asexual **7**:926, **7**:989
    sexual **7**:926–**7**:927
  rhizomes **7**:918, **7**:*919*,
    7:920–**7**:*921*, **7**:989
  roots **7**:918–**7**:920,
    7:924–**7**:925
  "seed" **7**:*918*, **7**:*919*
  stems and leaves **7**:918,
    7:*919*, **7**:923–**7**:*924*
  sweet **7**:917, **7**:922
  tubers **7**:918, **7**:*919*,
    7:921–**7**:922,
    7:*923*, **7**:*925*, **7**:926
  varieties **7**:926
potato blight **6**:770

potato family **7**:*916*,
  7:917
  poisons in **7**:*925*
prunus **1**:*36*
puffball **6**:771
  giant **6**:778
  wolf-fart **6**:*778*

Q

quince **1**:*36*

R

rainbow plant **10**:1301,
  10:1303
raspberry **1**:37, **1**:38
redwood **8**:*1138*
  California/coastal
    8:1139, **8**:1142,
    8:1144, **8**:1145,
    8:1146
  dawn **8**:1140
  reproduction, asexual
    7:*988*, **7**:989
rice **6**:752
rose **1**:*36*, **1**:37, **1**:39
rosids **1**:*36*, **1**:37
rush **6**:*744*
rust fungi **6**:771

S

saguaro cactus
  8:**1048**–**8**:**1059**
  birds and **8**:*1057*
  "boots" **8**:*1057*
  external anatomy
    8:*1051*
  flowers **8**:*1051*,
    8:*1058*–**8**:1059
  fruit **8**:*1052*, **8**:1059
    roots **8**:1054
  internal anatomy
    8:*1055*, **8**:1056
  pollination **8**:*1058*
  seed dispersal **8**:1058
sargassum **8**:1126, **8**:1134
sea lettuce **8**:*1126*,
  8:*1129*, **8**:1135

sea lettuce *(continued)*
  reproduction
    8:1136–**8**:1137
seaweed **8**:**1126**–**8**:**1137**
  cenocytic **8**:1133
  external anatomy
    8:1127,
    8:*1128*–**8**:*1131*
  fossil **8**:1130
  internal anatomy
    8:1127,
    8:*1132*–**8**:*1133*
  niches **8**:1131
  reproductive system
    8:1127,
    8:*1134*–**8**:*1137*
sedge **6**:*744*
seed-bearing plants **1**:*36*,
    1:37; **6**:*744*, **6**:*824*;
    7:*916*, **7**:917;
    8:*1048*, **8**:*1138*;
    10:*1300*
sequoia **8**:**1138**–**8**:**1149**
  external anatomy
    8:1139,
    8:1140–**8**:*1143*
  internal anatomy
    8:1139,
    8:*1144*–**8**:*1145*
  reproductive system
    8:1139,
    8:1146–**8**:*1149*
slime mold **6**:770
smut fungi **6**:771
spike moss **2**:285
stinkhorn **6**:771, **6**:779
strawberry **1**:37, **1**:38,
    1:43, **1**:44; **7**:989
sugarcane **6**:744
sundew **10**:*1300*,
    10:1301,
    10:1303–**10**:1304,
    10:*1307*, **10**:*1312*
sunflower, seeds **6**:752
support structures
    9:*1161*

T
toadstools **6**:*771*
tobacco **7**:*916*
tomato **7**:*916*
trees
  evergreen **8**:1140
  rings and
    dendrochronology
    8:1145
truffle **6**:*774*, **6**:775,
    6:*775*
trumpet pitcher **10**:*1300*

V
vascular plants
    1:*36*–**1**:37; **2**:276;
    6:*744*, **6**:*824*;
    7:*916*–**7**:917;
    8:*1048*, **8**:*1138*;
    10:*1300*
Venus flytrap
    10:**1300**–**10**:**1313**
  digestive system
    10:1301,
    10:*1308*–**10**:*1311*
  external anatomy
    10:1301,
    10:1302–**10**:*1305*
  internal anatomy
    10:1301,
    10:*1306*–**10**:*1307*
  reproductive system
    10:1301,
    10:*1312*–**10**:*1313*

W
waterwheel plant
    10:1301, **10**:1305
wheat **6**:751, **6**:752
whisk fern **2**:*276*, **2**:282
wrack **8**:1127
  bladder **8**:*1126*,
    8:*1128*, **8**:1129,
    8:1131, **8**:*1135*,
    8:*1137*

Y
yeast **6**:776
  reproduction **6**:777
yew
  English **8**:1148
  plum **8**:1148

Index of biological systems

Circulatory system

A

altitudes, high, living at 4:566
amebocytes 1:138
aorta 1:115
arteries 1:*118*; 4:487
  hardening of the 4:567
arterioles 1:118, 119, 121
arteriosclerosis 4:567
atherosclerosis 1:*118*
atrioventricular (AV) node 1:116–117; 6:767
autoregulation 1:121

B

blood 1:*108–109*
  clotting 1:*119*
  sugar levels 2:*272*
blood cells 1:*108–109*
  platelets 1:108, *109, 119*
  red (erythrocytes) 1:*108–109,* 119; 4:*521*; 8:*1024, 1029*
  white (leukocytes) 1:*108, 109*; 5:611
blood pressure 1:*121*
bulbus arteriosus 8:1070, 1104, 1105; 9:1248

C

capillaries 1:119–*120*
cardiac ganglion 6:767
carotid arteries 1:121
chordae tendineae 1:*116*
circulatory system 1:**106–121**
  open and closed systems 1:107

regulation and control 1:*121*
  see also arteries; blood; capillaries; heart; veins
cloacal bursae 9:1275
clotting 1:*119*
conus arteriosus 4:456; 9:1248
coronary arteries 1:115
countercurrent heat exchangers 1:120
  in eagles 2:231
  in fish 8:1069
  in otters 6:856
  in penguins 7:*877–878*
  in seals 8:1124
  in whales 2:202; 3:391
ctenidia 6:*821*; 8:1017

D

diastole 1:117

E

erythrocytes *see* blood cells, red
erythropoietin 1:108

F

fibrin 1:119

H

heart
  of amphibians 1:*110,* 111–112
    bullfrogs 1:*61*
  of arthropods
    honeybees 1:117
    spiders 9:1261
  of birds 1:*110,* 112, 115
  of fish 1:*110*–111; 9:1248
    coelacanths 1:129
    hagfish 1:111, *112;* 4:*456*

sharks and rays 4:473; 9:1248
heartbeat 1:*114,* 115–117; 6:*766,* 767
  bats 3:301
  elephants 2:258, *259*
  humans 1:*113*–117; 4:*567*
  mammals 1:112; 4:487
  neurogenic 6:767
    chameleons 5:*632*
    crocodiles 1:*112;* 2:159
    snakes 3:407
    reptiles 1:*110,* 112
    tortoises 9:*1275,* 1276
    turtles 9:*1203*
  transplants 5:*617*
  valves 1:*113, 116*
  "venous hearts" 1:*115;* 3:307
  of vertebrates 1:*110–112*
    cardiac muscle 1:67, 115–116; 2:254; 6:755, *756, 766–767;* 7:*954;* 10:1338
heart attacks 1:115; 4:567
heat exchangers, countercurrent *see* countercurrent heat exchangers
hemocoel 3:332; 4:500; 5:684, 696
  of ants 1:30
  of centipedes 3:336
  of scorpions 8:1080
  of weevils 10:*1355*
hemocyanin 1:138; 5:688; 6:821; 9:1261

hemoglobin 1:108, *109,* 119; 4:456; 5:678; 8:1015, *1023–1025*
  albatross 1:17
  earthworm 2:241
  elephant 2:259
  fetal 8:1024
  otter 6:857
  penguin 7:878
  platypus 7:893
hemolymph 3:332; 5:684
  ant 1:30
  centipede 3:334, 336, 338
  dragonfly 2:213, 216, 217
  hawkmoth 4:500, *503*
  honeybees 4:535
  housefly 4:*544, 547,* 548
  ladybug 10:*1348*
  louse 5:696
  scorpion 8:1080
  spider 9:1261
  weevil 10:*1352,* 1353, *1355*
hepatic portal system 1:116
hepatic portal vein 5:588
hepatic sinus 8:1120

M

mitral valve 1:*116*
myoglobin 2:203; 6:761, 839, 857; 7:875, 876, 893, 973; 8:1024, 1067, 1117; 9:1286; 10:1427

O

ostia 5:688; 10:1355

P

pacemaker
  in decapods 6:767
  in vertebrates 6:767

Panizza, foramen of
2:159–160
pericardium 1:113
plasma 1:108; 5:618
platelets 1:108, 109, 119
processus entoglossus
5:630
prothrombin 1:119
pulmonary artery 1:113;
4:567
pulmonary valve 1:113
pulmonary veins 1:115
pulse 1:117

R
rete mirabile 1:120;
3:369, 391; 7:878

S
sinoatrial (SA) node
1:116; 6:767
sinus venosus 8:1070,
1104, 1105; 9:1248
stem cells 1:108
syrinx 10:1414
systole 1:117

T
thromboplastin 1:119

V
veins 1:118; 4:487
vena cava 8:1120–1121
ventricular fibrillation
1:117

Digestive system

A
abomasum 3:370; 7:979;
10:1376, 1377,
1380–1381
alcohol dehydrogenase
4:568

alimentary canal,
mammalian 6:741
ambergris 2:205
amylase 2:173; 4:569
antioxidants 1:72
appendix 1:102; 2:182;
4:569

B
baleen 2:175; 3:375, 377,
378, 392, 393
baleen plates 2:189
beak, octopus 2:175
bile 2:182, 204; 5:590;
6:741; 10:1399
Brunner's glands 2:182

C
cecal pellets and feces
4:490–491
cecum 1:102; 2:181, 182;
4:490, 569;
8:1045; 9:1220
of eagles 2:230
cellulose digestion 6:739;
10:1311
in cows 5:659
in deer 7:979–980
in even-toed ungulates
4:506–507
in giraffes 10:1377
in hares and rabbits
4:490–491
in kangaroos 5:659,
660; 10:1377
in mice 7:961
in porcupines
7:912–913
in rhinos 8:1045
in sirenians 5:714
in sloths 9:1190,
1191
in squirrels 9:1220
in wildebeests 10:1376,
1377
in zebras 10:1435

chelicerae 9:1253, 1254
scorpion 8:1077
spider 2:175; 9:1254,
1256, 1262
chlorogogen 2:184, 242
chyme 2:181–182, 204
cloaca
of amphibians
frogs 1:63
newts 6:810
of birds 5:589
albatrosses 1:18, 19
of mammals 2:185
marsupials 2:185
monotremes 2:185;
7:883, 894,
895–896
of reptiles
crocodiles 2:161, 162
turtles 9:1205
cnidae see nematocysts
coelenteron 5:637, 642,
644–645; 8:1088,
1089, 1091, 1093;
9:1166
colon 2:182
coprodeum 2:162
coxal glands 8:1081;
9:1263
crop
of birds 2:178, 230
of dragonflies 2:214
of hummingbirds 5:589
crop milk 7:879

D
defecation 2:186
diastema 3:427; 5:655;
7:905, 953; 9:1215
digestive and excretory
systems 2:172–187
invertebrate
excretory systems
2:187
intestines 2:183
mouthparts 2:175

vertebrate excretory
systems 2:186–187
intestines
2:180–183; 7:943
liver, pancreas, and
2:184
lymphoid tissue
2:181
mammalian 6:741
stomach 2:177–178
teeth 2:173–174
duodenum 2:181, 182,
204; 4:568, 569;
10:1399

E
epiglottis 2:176
esophagus 2:176
excretory systems see
digestive and
excretory systems

F
fascicles 5:697; 6:763,
764
fat body 10:1353
feces 2:185–186
elephant dung 2:261

G
gallbladder 2:184; 4:568
missing in giraffes
3:371
gastroliths
in birds 2:178; 7:879
in crocodiles 2:161
gills, of giant clams
3:345, 347, 352
gizzard
of birds 2:178
albatrosses 1:18
eagles 2:230
woodpeckers
10:1415
of crocodiles 2:161
of invertebrates 2:179

glomeruli **8**:1045
glossa **4**:530
glycogen **5**:680;
    8:1106–1107
goblet cells **4**:568

H

haustellum **5**:697
hepatopancreas
    8:1106–1107
homodontia **1**:62
hydathodes **6**:748

I

ileocecal valve **9**:1277
ileum
    insect **2**:183
    toothed whales **2**:205
    vertebrate **2**:181, 182
intestines **2**:180–183;
    7:*942*
    and diet **3**:325
    of eagles **2**:230
    of toothed whales
      2:204, 205

J

jejunum **2**:181, 182

K

kidneys **2**:185, 186;
    8:1045
    lobular/reticulate
      8:1123

L

labellum (of insects)
    houseflies **4**:*541*
    mosquitoes **4**:*546*
labial palps **3**:*352*
labium **4**:540–541; **5**:697
labrum **4**:540–541;
    5:697
larynx **2**:176; **10**:1414
Lieberkühn, crypts of
    4:*569*
lips **2**:173

liver **1**:116; **2**:182, 184;
    5:680; **8**:*1045*,
    1106–1107

M

Malpighian tubules
    2:187; **4**:536
    of ants **1**:33
    of centipedes **3**:336,
      340
    of houseflies **4**:548
    of lice **5**:*697*, 698
    of spiders **9**:1257, 1263
    of weevils **10**:1353
mandibles **1**:135; **5**:684,
    692, 693, 697
    of beetles **10**:1348
    of caterpillars **4**:499
    of centipedes **3**:339
    of weevils **10**:1359
mastication **2**:173
maxillae **5**:684, 697
    of centipedes **3**:339
    of lobsters **5**:689
    of spiders **9**:1254
maxillipeds **5**:685
    of centipedes **3**:333,
      339, *340*
    of crabs **1**:135, *139*
    of lobsters **5**:689
microvilli **2**:179, *180*,
    182, 183
mycetome **5**:697

N

nematocysts (cnidae)
    5:637, *640*; **8**:1086,
    1092–1093
nephridia **2**:187
    of earthworms **2**:237,
      242

O

omasum **7**:980; **10**:1375,
    1376
oral (buccal) cavity **2**:173

P

palate, human **2**:*173*
pancreas, digestive
    function **2**:184;
    5:589; **6**:741;
    8:1107
pedipalps **9**:1253
    scorpion **8**:1075, *1076*,
      1077, *1078*
    spider **2**:175; **9**:1262
peristalsis **2**:178; **3**:*371*
peristomium, of
    earthworms **2**:234,
    236, 240, 242
peritrophic envelope
    2:214
pharyngeal spines, turtle
    2:*176*
pharyngeal teeth,
    sandworm **2**:178
pharynx **2**:176
proboscis
    of butterflies and
      moths **4**:*497*
    of houseflies **4**:541
proctodeum **1**:18; **2**:162
prostomium, of
    earthworms **2**:240
proventriculus **1**:18, 32;
    2:230; **5**:589; **7**:879;
    10:1415
pyloric ceca **9**:1291
pyloric sphincter
    (pyloric valve)
    9:1220, 1277
pylorus **2**:183

R

radula **2**:175, 179
    octopus **2**:*175*; **6**:*817*,
      822
rectal glands **1**:130;
    9:*1234*, 1250
rectum
    insect **2**:183
    vertebrate **2**:182

reticulum **2**:177; **3**:*370*;
    7:980; **10**:1375,
    1376
rumen **2**:177; **3**:*370*,
    371; **7**:980;
    10:1375–1376,
    1377

S

saliva **2**:173, 274
scats **3**:426
stercoral pocket **9**:1257
stomach **2**:177–178
    of ants **1**:32
    cardiac **1**:139; **5**:689;
      9:1234, *1235*
    of carnivores **3**:325
    of cows **5**:*659*
    of crabs **1**:139
    of deer **7**:*979*, 980
    of eagles **2**:230
    of kangaroos
      5:*659*–660
    of lobsters **5**:689
    pyloric **1**:139; **5**:689;
      9:1234
    of ruminants **2**:177;
      7:979
    of tarantulas **9**:*1258*,
      1262
style (animal) **2**:179
    crystalline **3**:352
stylets, of lice **5**:697,
    698

T

teeth **2**:173–175
    canine **2**:174;
      3:418–419; **4**:*515*;
      5:678; **6**:*851*
    of chimpanzees **1**:*96*,
      101, *103*
    of deer **7**:*969*, 971
    of horses **10**:1435
    of mandrills **6**:*732*,
      739, *740*, 741

teeth (continued)
of wolves **10**:1391,
1398, 1399
carnassial **3**:419; **5**:596,
665; **6**:847, 851;
7:942; **10**:1332,
1398
diphyodont and
polyphyodont
5:713
incisor **2**:174; **5**:655;
6:741, 851; **7**:905,
950, 971; **8**:1044;
9:1215; **10**:1399
lingual **4**:450
mammal **2**:173, 174
development **6**:740
maxillary **9**:1291
molars **2**:173, 174;
10:1435
pharyngeal **2**:178
premolars **2**:174; **5**:596;
8:1044; **10**:1435
vomerine **9**:1291
see also tusks
trimethylamine oxide
(TMAO) **1**:130
tusks **2**:174
elephant **2**:247–248

U

urea **1**:62; **2**:186; **3**:427;
8:1123
uric acid **2**:186; **3**:408;
9:1277
urodeum **2**:162

V

villi **2**:179, 180, 182;
4:569; **6**:741
vitamins **1**:72
vitamin B **4**:489

W

water loss, prevention, by
albatrosses **1**:17

Endocrine and exocrene systems

A

ACTH
(adenocorticotropic
hormone) **2**:271
adrenal glands **2**:268,
271–272; **4**:570
adrenaline
see epinephrine
amylase **2**:173; **4**:569
anal glands
of pumas **7**:933
of weasels **10**:1335,
1336
of wolves **10**:1399
antidiuretic hormone
(ADH) **2**:268;
4:571
auxin **2**:266

C

calcitonin **2**:268
catecholamines
2:271–272
caudal glands **10**:1389
cholecystokinin **2**:272
cortisol (hydrocortisone)
2:272

D

diabetes **4**:571; **5**:622
digestive glands **2**:172
fundic glands **2**:177
Dufour's gland **1**:30;
4:532–533

E

ecdysone **2**:267
endocrine and exocrine
systems **2**:**264–275**;
6:791

endocrine system
2:264–265
vertebrate **2**:265,
270–273
see also hormones
exocrine system **2**:264,
265, 274–275
skin glands **2**:265;
9:1218
epinephrine **1**:121;
2:271–272
estrogen cycle **2**:269
estrogens **7**:998–999
ethylene, plant
hormone **2**:266

F

follicle-stimulating
hormone (FSH)
2:271; **7**:997–998,
999

G

gastrin **2**:272
GHRH **2**:270
gibberellins **2**:266
glucagon **2**:272, 273;
4:571
glucocorticoids **2**:272
gonadotropin-releasing
hormone (GnRH)
7:997–998; **10**:1416
gonadotropins **2**:269,
271

H

hormones **2**:264
affecting growth and
development **2**:265,
266–267
fungal **2**:269
gastrointestinal (GI)
2:272
half-life **2**:271
and homeostasis **2**:265,
268

lipid-insoluble **2**:264,
265
lipid-soluble **2**:265,
271
mode of action **2**:264,
265
plant **2**:266
and reproduction
2:265, 269
eagle **2**:233
sex, of vertebrates
7:996–999
types **2**:265
hypocalcin **2**:268
hypopharyngeal gland
4:532
hypopharynx **5**:697
hypophysis **2**:266
hypothalamus **2**:270;
4:565; **5**:634
effect on the pituitary
2:266, 270; **4**:571;
6:791; **7**:997–998
of manatees **5**:709

I

insulin **1**:70; **2**:272, 273;
4:571; **9**:1291
from genetic
engineering **1**:50

J

juvenile hormone (JH)
2:267; **4**:505

L

lachrymal glands **2**:275
Langerhans, islets of
4:571
luciferase **10**:1349
luciferins **3**:439; **10**:1349
luteinizing hormone
(LH) **2**:269, 271;
4:571; **7**:998, 999

M

mammary glands 2:*275*
   human 2:*275*; 4:571
   ungulate 2:*275*
   whale 3:394, *395*
mandibular gland 1:30;
   4:532
melanophore-stimulating
   hormone (MSH)
   5:631
melatonin 2:273; 9:1202
metapleural gland 1:30
mineralocorticoids 2:272
mucus 5:609
mucus glands 2:274
musk glands, crocodile
   2:152, 161

N

neurohormones 2:270,
   275
norepinephrine 1:121;
   2:272

O

oxytocin 2:271

P

pancreas, endocrine
   function and
   hormones 1:70;
   2:*272*, 273; 4:571;
   8:1107
parathyroid gland 2:268
   of eagles 2:233
parietal organ 1:60
pedal glands 8:1047;
   10:1363, 1373
pharyngeal
   ultimobranchial
   glands 2:268
pheromones
   crab 1:140
   elephant 2:263
   of lions 5:676
   porcupine 7:908

scorpion 8:1082
   as sexual attractants
      2:274
   social insect 2:275
      ants 1:30, 31
   as territory markers
      2:274
   triggering ovulation,
      porcupines 7:914
pineal gland (pineal
   body; parietal eye;
   third eye) 1:60;
   2:273; 9:1202, *1274*
pituitary gland 4:*571*
   hormones 2:*266*, 268,
      270–271;
      7:997–998
poison glands *see* venom
   glands and venom
preen gland (uropygial
   gland) 1:13;
   10:1409
preorbital glands 10:1363
progesterone 2:269;
   7:*995*, 999
prolactin 2:271, *273*
prothoracic glands,
   arthropod 2:267

R

repugnatorial glands
   3:340
royal jelly 4:532

S

salivary glands 2:274
   of anteaters
      3:325–326
   of centipedes 3:336
   of humans 4:570
   of lice 5:697, *698*
salt glands
   albatross 1:*18*
   crocodile 2:*152*, 161
   marsh grass 6:748
   penguin 7:879

sebaceous glands, puma
   7:933
secretin 2:272
serous glands 2:274
somatostatin 2:267, 270
somatotropin 2:266, 270
stannius gland 2:268
steroids 2:265, *271*
sublingual glands 2:274
suborbital glands 7:976
sweat glands 2:274–275

T

T3 (thyroid hormone)
   2:267, 271
T4 (thyroid hormone)
   2:267, 271
testosterone 4:571;
   7:992, 995, 998
   and birdsong 7:999
   in hyenas 5:601, 604
   in mandrills 6:*730*, 743
thyroid gland 2:267
thyroid hormone 2:267
thyroid-releasing
   hormone (TRH)
   2:267
thyroid-stimulating
   hormone (TSH)
   2:267, 270–271
thyroxine 2:267

U

urophysis gland 2:267
uropygial gland (preen
   gland) 1:13;
   10:1409

V

vasopressin 1:121
venom glands and
   venom
   of Cane toads 2:275
   of centipedes 3:333,
      336, 339, *340*
   of honeybees 4:533

of jellyfish, coral, and
   sea anemones 5:640
of newts and
   salamanders 6:804
of octopuses 6:822
of scorpions 8:1081
of snakes 3:404
of spiders 9:1254,
   1256, 1263
of stingrays 9:*1241*
venoms that attack
   nervous system
   6:783

W

wax glands 4:533

Immune system

A

Addison's disease 5:623
allergies 5:*613*, 615–616
anaphylatoxins 5:611
ankylosing spondylitis
   5:622
antibodies 5:608, 609,
   612–614, 620
   autoantibodies 5:622
   see also
      immunoglobulins
antibody-dependent
   cell-mediated
   cytotoxicity
   (ADCC) 5:613, 615
antigenic drift 5:621
antigenic shift 5:621
antigens 5:612, *613*
arthritis, rheumatoid
   5:622, *623*
attacins 5:608
autoimmunity 5:608,
   622, *623*

B

basophils **1**:*108*; **5**:611
B-cells *see* lymphocytes
"bubble boy disorder"
    (severe combined
    immunodeficiency)
    5:*622*, 623

C

cancer **1**:*86*, 87; **10**:1323
cecropins **5**:608
Chagas' disease **5**:622
clonal selection **5**:620
complement cascade
    5:609, 611, 614

D

dendritic cells **5**:611

E

eosinophils **1**:*108*, 109;
    5:611

G

genetic disorders,
    immune disorders
    5:*622*, 623

H

hay fever **5**:*613*
histamine **5**:*613*
hypersensitivity,
    immediate **5**:616

I

immune, defense, and
    lymphatic systems
    5:**606–623**
immune disorders
    5:*622–623*
immunity **5**:618,
    621
    adaptive **5**:606, 608,
        612–617
    innate **5**:606, 608,
        609–611

lymphatic system
    4:488; **5**:606,
    618–621
    see also autoimmunity
immunoglobulins (Igs)
    5:612, 613
IgE **5**:615
    surface **5**:615, 620
    see also antibodies
inflammation **5**:606–608,
    611
influenza, pandemics
    5:621
interferon **5**:609–610,
    614

L

leukocytes **5**:611
lymph **4**:488; **5**:608, 618,
    619
lymphatic system *see*
    immune, defense, and
    lymphatic systems
lymph nodes ("glands")
    5:618, *619*
lymphocytes **1**:*108*, 109;
    4:488; **5**:608, 621
    B (B-cells) **5**:612, *615*,
        617, *618*–619,
        620–621
    CD4 **5**:*616*; **10**:*1315*,
        1323
    helper T-cells **5**:616,
        617, 619
    production of
        5:*618*–619
    T (T-cells) **5**:*615*, *616*,
        617, *618*, 619
lysozyme **5**:608, 611

M

macrophages **5**:*610*, 611
major histocompatibility
    complex (MHC)
    5:617, 618, 622
malaria **5**:*607*

mannose binding lectin
    (MBL) **5**:611
mast cells **5**:611, *615*
membrane attack
    complex **5**:611
MHC *see* major
    histocompatibility
    complex
monocytes **1**:*108*, **5**:*609*

N

neutrophils **1**:*108*, 109;
    5:608, *610*, 611

O

opsonization **5**:611, 613

P

pathogens **5**:606
phagosomes **5**:611
pneumonia **5**:*607*
pus **5**:608, 611

R

rheumatic fever,
    streptococcal **5**:622

S

self-tolerance **5**:620
severe combined
    immunodeficiency
    (SCID) **5**:*622*, 623
shock, anaphylactic **5**:616
systemic lupus
    erythematosus
    5:623

T

T-cell receptors (TCRs)
    5:608, 617, 620
T-cells *see* lymphocytes
tears **5**:611
tetanus (lockjaw) **5**:*607*
thymocytes **5**:619
thymus **5**:619, *621*
thyroiditis **5**:622

toll-like receptors
    (TLRs) **5**:610–611
transplants, organ **5**:*617*
trypanosomiasis,
    American **5**:622
tuberculosis (TB) **5**:606
typhoid **5**:*607*
typhus **5**:*607*

W

Wiskott-Aldrich
    syndrome
    (WAS) **5**:623

Muscular system

A

abductor muscles, of
    coelacanths **1**:127
actin filaments **6**:754,
    756, 757–*759*, 760,
    767, 768
adductor muscles
    of clams **3**:349, *350*
    of coelacanths **1**:127
ambulacral muscles
    9:*1231*

B

biceps femoris
    elephant **2**:255
    human **4**:561
    rhino **8**:1041
    seal **8**:1111
biceps muscles **4**:*560*;
    6:734, *754*; **8**:1040

C

calmodulin **6**:769
cardiac muscle **1**:67,
    115–116; **2**:254;
    6:755, 756,
    766–767; **7**:954;
    10:1338

D

deltoid muscles **1**:15
depressor caudae muscle
    2:228
diaphragmaticus muscle,
    crocodile **2**:160;
    8:1028
dorsoventral muscles
    6:*762*

E

enzymes, digestive
    10:1339–1340
epaxial muscles **6**:853
erector pili muscles
    7:937
erector spinae muscles
    4:561
expansor secondarium
    muscles **2**:227–228
extensor carpi radialis
    1:*15*
extensor carpi ulnaris
    1:*15*

F

fast- and slow-twitch
    7:973–974; **9**:1201
flexor carpi ulnaris **1**:*15*

G

gastrocnemius muscle
    4:561; **6**:839;
    10:1412
geniohyoid muscles
    10:*1412*
gluteal muscles **4**:561
gluteus maximus **4**:561
gluteus medius **4**:561
gracilis muscles **8**:1111

H

hamstring muscles **4**:561
hemipenis **5**:*634*; **7**:*997*
hyoglossus muscle **5**:633
hyoglossus tendon **5**:*630*

I

iliocostal (iliocostalis)
    muscles **2**:156;
    8:1117
intercalated disks
    6:766
intercostal muscles
    10:1412
intertransversales muscle
    2:228
intramandibularis muscle,
    crocodile **2**:156

L

lateral masseter muscle
    9:*1216*
latissimus dorsi muscles
    1:15; **4**:561; **8**:1041;
    10:1370
levator caudae muscle
    2:228
levator coccygeus muscle
    2:228
ligamentum nuchae
    3:364
longissimus caudalis
    muscles, crocodile
    2:157
longissimus muscles
    7:936–937; **8**:1117
longus colli **10**:1412

M

mandibular depressor
    muscle, crocodile
    2:156
masseter muscles **6**:853;
    7:*954, 955*; **9**:1216,
    1217; **10**:*1334*
multifidis cervicis muscle
    2:228
muscular system
    6:**754–769**
    antagonistic muscles
        6:734, *754*
    circular muscles **6**:*755*

force versus speed
    6:764
mode of action **6**:754,
    756–761
muscle cells (myocytes)
    1:66–67
    energy for **6**:761
    red and white muscles
        8:1012, 1067;
        9:*1286*
mustelids **6**:*844*, 845
mylohyoid muscles
    3:*383*
myocytes **1**:66–67
myofibrils **6**:757
myomeres
    of hagfish **4**:452
    of sailfish **8**:*1066*
myosepta **1**:127
myosin filaments **5**:672;
    6:754, 756–*759*,
    768
myotomes **1**:127; **9**:1285,
    1286

N

neuromuscular junctions
    6:760–*761, 791*

O

orbicularis oris muscle
    6:*764*

P

pallial muscles, of clams
    3:*350*
pectoralis major muscles
    1:15; **6**:*839*; **7**:875
pectoralis minor muscles
    1:15
pectoralis muscle **4**:561
    of bats **3**:302
    of birds **1**:15; **3**:302;
        10:1412
processus entoglossus
    5:633

pterygoid muscles
    9:1217

R

rectus abdominalis **7**:936
retractor muscles, of
    clams **3**:*350*–351
rhomboideus muscle
    8:1040–1041;
    10:1370
rigor mortis **6**:759

S

sarcolemma **6**:757
sarcomeres **6**:756, 757,
    759
sarcoplasmic reticulum
    6:757, 760
scapulohumerus muscles
    1:15
semispinalis capitis
    10:1412
semitendinosus muscles
    8:1117
serratus anterior muscles
    10:1412
serratus ventralis muscles
    10:1370
skeletal muscles **1**:66–67;
    2:254; **6**:755,
    756–757, 762–765;
    10:1392
smooth muscles **1**:67;
    2:254; **6**:755, *756,
    768–769*; **7**:890,
    954
soleus muscle **4**:561
sphincter muscles **6**:763,
    764
striated/striped muscles
    5:630; **7**:890, 954;
    10:1392
superficial masseter
    muscle **9**:1216
superior rectus muscle
    8:1069

supracoracoideus muscles
1:15; 6:*839*; 7:875;
10:1412
supracoracoid muscles
5:586

T

temporal (temporalis)
muscles 9:1217;
10:*1334*
tendons 3:420; 6:763;
7:954
in a kangaroo's hind
legs 5:656
thermogenesis,
nonshivering 7:875
tongue
of chameleons 5:*630,
633*
of giant anteaters
3:*322–323*, 324,
325
of giraffes 3:*359*,
364–365
trapezius muscle 4:561;
8:1040–1041;
10:1370
triceps muscles 4:*560*;
6:734, *754*; 8:1040
of birds 1:*15*
tropomyosin 6:759–760
troponin complex 6:759

Nervous system

A

acetylcholine 1:121;
4:565; 6:760–761
action potentials
6:783–785, 786;
10:1310
ampullae of Lorenzini
4:470, *471*; 6:*787*;
9:*1247*

amygdala 4:565;
5:709;6:798
axons 6:782, *785*
giant, of squid 6:783

B

balance, sense of 6:787,
788–789
blood-brain barrier
6:795
brachial plexus 5:587
brain 6:781, *794*, *795*,
796–799
imaging the 6:*799*

C

central nervous system
(CNS) 4:*562*;
6:*780*, 794,
796–799; 7:938,
956; 9:1259
and the endocrine
system 2:264; 4:563
invertebrate 6:*792–793*
mode of action 6:781,
782–785
central pattern
generators 6:795
cerebellum 6:*797*
of chimpanzees 1:98,
99
of dolphins 2:199
of elephants 2:257
of humans 4:*564*;
6:*794*, *795*
of hummingbirds 5:587
of manatees 5:709
of ostriches 6:840
of seals 8:1118
cerebral cortex 6:*798*,
799
of dolphins 2:199–200
of elephants 2:257
of humans 4:564;
6:*796*
of ostriches 6:840

cerebral ganglion, of
lobster 5:*690*
cerebrospinal fluid 4:564
cerebrum 6:797,
798–799
of birds 2:229
hummingbirds
5:587
ostriches 6:840
of dolphins 2:199–200
of elephants 2:257
of humans 4:*564*;
6:*794*
of mandrills 6:737
of pumas 7:*939*
of whales 3:385
of wolves 10:1395
chemoreceptors 6:790
housefly 4:546
spider 9:1260
circumesophageal
connectives 2:215
cochlea 10:1413
columella 1:60
cornea 4:563
flat 7:876
corpora allatum 2:215
corpora cardiaca 2:215
corpus callosum 5:657
cribriform plate 3:305
crossbreeds, deer 7:981

D

dementia, senile 6:*799*
dendrites 6:782
deutocerebrum 3:337
diencephalon 10:1395

E

ear, human 4:*565*; 6:*789*
echolocation
by fruit bats 3:305
by whales 3:375, 388
electric organs, in
electric rays
9:*1246*

electroreceptors 6:790
in coelacanths 1:123,
124–125
in echidnas 7:892
in platypuses 6:*790*;
7:891–*892*
in sharks and rays
4:470; 6:*787*;
9:*1247*
encephalization quotient
(EQ) 4:564
eyes
cat 5:*668*, 676; 7:*939*
cephalopod 6:788
compound 1:*137*;
4:496, 502, *539*
human 4:*563*; 6:*788*
lobsters 5:686–687
muscles 6:*765*
octopus 6:*788*, 820
shark 4:469–470, *471*
simple
of giant clams
3:*353*
of scorpions 8:1079
see also ocelli
snake 3:*399*, 400, 406
superposition 4:502
vertebrate 6:788
whale 3:*387*
eyespots, of starfish
9:1232

F

fasciculus aberrans 5:657
fovea 2:229; 8:1103;
9:1218

G

ganglia 6:780–781, 792,
792–793
of bivalves 3:*353*
of earthworms
6:792–793
of giant centipedes
3:*337*

of giant clams **6**:*792*
of hawkmoths **4**:*502*
of houseflies **4**:*545*
of humans **4**:563–564
of scorpions **8**:*1079*
of tarantulas **9**:1257, *1259*
of weevils **10**:1354
glial cells **6**:785
gyri **4**:565; **6**:*737*

H
hearing **6**:*789*
hippocampus **4**:565; **6**:798
hypothalamus **6**:798

I
intelligence, relative **4**:564
interbreeding, deer **7**:981
interneurons **2**:256; **3**:366; **6**:*783*; **7**:908; **8**:1042; **10**:1336

J
Jacobson's organ *see* vomeronasal organ

L
lateral line **1**:125, *128*; **3**:*442*; **4**:469; **6**:*790*; **9**:*1287*, *1288*
of newts **6**:*807*
of octopuses **6**:820
of rays **9**:1247
of sailfish **8**:1068, *1069*
limbic system **6**:798

M
magnetic fields, detection **6**:791; **9**:1202

mechanoreceptors **6**:787, 788–789
earthworm **2**:240
housefly **4**:546
spider **6**:*791*; **9**:1259–1260
medulla oblongata **6**:796
of chimpanzees **1**:98
of dolphins **2**:199
of pumas **7**:939
of trout **9**:1287
of wolves **10**:1394–1395
meninges **7**:938
mesencephalon **10**:1395
metencephalon **10**:1395
myelencephalon *see* medulla oblongata
myelin sheath **4**:453–454; **8**:*1118*

N
nerve cells *see* neurons
nerve net **6**:792
of jellyfish **5**:637, 641
of sea anemones **8**:*1090*
nervous system **6**:**780–799**
neuromasts **3**:442; **9**:*1287*, *1288*
neuromuscular junctions **6**:*760–761*, *791*
neurons **1**:66; **3**:366; **4**:502; **6**:768, 781, 782–785; **7**:909
giant, of cephalopods **6**:783, *820*
named **6**:793
types
interneurons **2**:256; **4**:562; **6**:782, *783*; **8**:1042
motor **2**:256; **4**:562; **6**:782, *783*; **7**:956; **8**:1042

sensory **2**:256; **4**:562; **6**:782, *783*; **8**:1042
neurotransmitters **4**:562; **6**:*769*
see also acetylcholine
nociceptors **6**:789
notocord **7**:1004

O
ocelli
of ants **1**:31
of honeybees **4**:534
of jellyfish **5**:643
of myriapods **3**:337
of starfish **9**:1232
olfactory epithelium **1**:16; **3**:304–305
ommatidia **4**:534
of dragonflies **2**:215
of houseflies **4**:545–546
operculum **1**:60

P
paleopallium **4**:564–565
pectines **8**:1077, 1082
peripheral nervous system (PNS) **4**:562; **6**:*780*, 794; **7**:938, *956*
autonomic nervous system (ANS) **4**:563; **6**:*766*, 794, 795; **7**:938
of hummingbirds **5**:587
somatic nervous system **4**:563; **6**:794–795
photoreceptors **6**:789
extraretinal **9**:1202
pupils, of cats **7**:*939*
Purkinje nerve cells **6**:*797*

R
receptor cells **6**:786
reflexes **6**:796
retina **4**:563
of albatrosses **1**:16
of chimpanzees **1**:99
of eagles **2**:229
of fruit bats **3**:304
of humans **6**:*786*, 788
of lions **5**:675, *676*
of mammals **6**:*791*
of ostriches **6**:840
of sailfish **8**:1069
rhabdomeres **9**:1260
rhinarium **6**:737
rhopalia **5**:*643*; **6**:789
rostral organ **1**:125, *128*

S
Schwann cells **6**:*785*
semicircular canals **9**:1288
sensilla **4**:497
sensory adaptation **6**:786–787
sensory and motor systems **6**:781, *786–791*
sensory organs **6**:787
sensory transduction **6**:787
setae **1**:137
smell, sense of **6**:787, 790
somatic nervous system **4**:563
spinal cord, human **6**:*796*
statocysts **1**:137; **6**:820
statoliths **5**:643
stereocilia **6**:*789*
subesophageal ganglion **2**:215
sulci **4**:565
synapses **4**:562; **6**:784–785, *791*

T

tapetum lucidum **1**:124, 128; **5**:676, 710; **6**:855; **8**:1118; **9**:1245
tarsal organs **9**:1260
thalamus **4**:565; **6**:797–798
thermoreceptors, tarantula **9**:1259, 1260
Tömösvary organ **3**:333
trichobothria **6**:*791*; **8**:1077, 1079; **9**:*1260*
tritocerebrum **2**:215
tympanic membranes **6**:790

V

vertebrate nervous system **6**:*794–799*
development in the embryo **7**:1004–1005
vision
  bat **3**:297, 303–*304*, *305*
  color **6**:*786*
  dichromatic **9**:1218
    and frugivory **6**:740
  seahorses and **8**:1103
  night-hunting snakes **3**:399
  stereoscopic (binocular) **1**:99; **4**:554; **5**:675; **6**:*736*; **9**:1187; **10**:1396
  underwater, penguins **7**:876
vomeronasal organ (Jacobson's organ) **3**:*406*; **4**:519; **5**:676; **6**:737; **7**:*976*; **10**:*1373, 1430*

W

whiskers
  of lions **5**:674
  of pumas **7**:*933*

Reproductive system

A

aedeagus **5**:699
  of dragonflies **2**:213, 218, 219
  of weevils **10**:1357
alternation of generations
  ferns **2**:*283–284*; **7**:*993*; **8**:1137
  hydrozoans **7**:986
  jellyfish **5**:646; **7**:986–987
  seaweeds **8**:*1134*–1135, 1137
  water fleas **7**:*989*
altricial young **4**:493
anthers **1**:43; **6**:751; **7**:*927*
apomixis **7**:*988*, 989
archegonia **2**:283, 284, 285
asexual reproduction **7**:985, *986–989*
  apomixis **7**:*988*, 989
  budding **7**:988
  parthenogenesis **7**:*986*, 988–*989*
  pedal laceration **8**:1094, *1095*
  polyembryony **7**:986
  regeneration **7**:988
  vegetative **7**:989
  see also fission
awns **6**:751, 753

B

baculum **7**:929; **8**:*1124*; **10**:*1341*, 1342, 1383, 1400
binary fission **1**:24, 53, 85; **7**:986
  longitudinal **8**:1094
blastocoel **7**:1004
blastocysts **8**:1124
blastula **7**:1004
breeding aggregates **3**:409
brood patch **1**:19
brood pouch **8**:1108, 1109
budding **7**:988

C

calyx **1**:43
carpels **7**:*1001*; **8**:1059
carpogonia **8**:1136
carpospores **8**:*1134*
caryopsis **6**:752
cervix **7**:996
chrysalises **4**:495
  butterfly **4**:504
clitoris **7**:996
cloaca **3**:326, 330
  of snakes **3**:408
colostrum **10**:1380
conceptacles **8**:*1136*, *1137*
cones
  giant sequoia **8**:*1140*, *1143*, 1146–1147
  larch **8**:*1143*
  pine tree **7**:*1000*, 1003; **8**:*1146*, *1147*
corolla **1**:43
corpus luteum **7**:*995*, 999

D

dioecious animals, centipedes **3**:336
dioecious plants **7**:1002

E

eggs
  bird **10**:1417
    albatross **1**:19
    ostrich **6**:*843*
  incubation **5**:591
  reptile **5**:635
    alligator or crocodile **2**:162, *163*
    python **3**:405
    tortoise **9**:1278–*1279*
    turtle **9**:1278
embryo, in animals **7**:*1004–1005*
  bone and skull development in vertebrates **9**:*1169*, 1170–1171
endometrium **7**:999
endosperm **6**:752–753
epididymis **4**:*572*; **7**:995
estrus cycle, of kangaroos **5**:663

F

fallopian tubes **4**:*572*
fertilization **7**:990–991
  double **7**:1002, *1003*
  external **3**:445; **7**:*992*, 993
  in flowering plants **7**:1002, *1003*; **10**:1313
  internal **7**:993
fission
  binary *see* binary fission
  longitudinal **8**:1094
  multiple **1**:24–25
flowers
  apple tree **1**:*39, 43*; **7**:*1001*
  cactus **8**:*1051*, *1058–1059*
  grass **6**:751–*752*; **7**:*1001*

maize **6**:*753*
orchid **6**:*825, 826, 828, 832–833*
perfect and imperfect **7**:1002
potato **7**:*921, 926, 927*
Venus flytrap **10**:*1312–1313*

G

gametangia **8**:1137
gametes (sex cells) **1**:67, *87*; **7**:985
  meiosis and **1**:85; **7**:*991*
gametophyte **7**:993, 1000, 1002
  fern **2**:277, 283, 284, 285; **7**:*993*
  seaweed **8**:*1134*
gas exchange **8**:*1012*, 1013–1015
gastrulation **7**:1004
gestation **7**:1005
  period, variable **3**:330
glumes **6**:751
gonopore **1**:140; **3**:341
grafting, apple trees **1**:45
granulosa cells **7**:995

H

hectocotylus **6**:*823*
hermaphroditism **7**:*998*
  sea anemones and **8**:1095
hymen **7**:996
hypanthium **1**:43

I

implantation, delayed **6**:860; **10**:1342
inbreeding, by grizzly bears **3**:429
ischial callosities **6**:*730*

K

karyogamy **6**:777–778
katoikogenic development **8**:1083

L

labia major **7**:996
labia minor **7**:996
larvae
  ammocoete **4**:448
  bipinnaria **9**:1237
  brachiolaria **9**:*1236*, 1237
  of crabs **1**:141
  of honeybees **4**:536
  of houseflies **4**:547, 548, *549*
  leptocephalus **3**:445
  of lice **5**:699
  of newts and salamanders **6**:*804*
  planula **5**:*646–647*; **7**:*986*; **8**:*1094*–1095
  of sea anemones **8**:*1094*–1095
  of starfish **9**:*1236*, 1237
  trochophore **3**:355
  veliger **3**:355
  zoea **1**:141
lemma **6**:751, *752*, 753
Leydig cells **7**:995
longitudinal fission **8**:1094

M

mammary glands
  dolphin **2**:207
mandrill **6**:*743*
megagametocytes **7**:1001
megaspores **7**:1001
menstruation **6**:743; **7**:999
micropyle **8**:1147
microsporangia **7**:*1000*; **8**:1146

midwives **2**:207
milk **2**:275
  production **2**:271, *273*
  seal **8**:1125
  whale **2**:207
monoecious plants **7**:1002
monospores **8**:1136
multiple fission **1**:24–25

N

Needham's sac **6**:823

O

oocytes, primary **7**:992
ova (eggs) **7**:*990*, 999
  development in humans **7**:*995*
  see also eggs
ovaries
  albatross **1**:19
  apple tree **1**:*43*
  chameleon **5**:634
  eagle **2**:232
  mammalian **7**:995
ovarioles **10**:1357
oviducts **7**:996
ovipositor **1**:35
  of dragonflies **2**:213, 218
  as a stinger **1**:35; **4**:533
  of weevils **10**:1359
ovoviviparity **3**:409
  in coelacanths **1**:131
  in sheep botflies **4**:549
ovulation **7**:*995*
  delayed **10**:1342
ovules **8**:1146–1147

P

palea **6**:751, *752*, 753
panicles **6**:751
parthenogenesis **7**:*986*, 988–*989*
pedal laceration **8**:1094, *1095*

penis
  of armadillos **3**:323
  of echidnas **7**:*895*
  of humans **4**:*572*; **7**:995
  of hyenas **5**:605
  of platypuses **7**:*895*
  of pumas **7**:*944*–945
  of reptiles **9**:1206
  of rhinos **8**:1047
  of whales **3**:383, *394*
pistils **1**:43
placenta (afterbirth) **3**:330, 356; **7**:1005; **10**:1379
  eating the **10**:1381
  hemochorial **6**:743
plasmogamy **6**:777
pollen **2**:276–277; **7**:1001, 1002
pollination
  of apple trees **1**:43, *44*
  by bats **7**:*1002*; **8**:*1058*
  of cacti **8**:*1058*
  of giant sequoias **8**:1146–1147
  of grasses **6**:751
  of orchids **6**:832, *833*
  of Venus flytraps **1**:131
polyembryony **7**:986
precocious young **3**:331, 373; **4**:493; **6**:843
promiscuity **4**:573
prostate gland **7**:995
pseudopenis, of hyenas **5**:602, 603, *604*
pupae **4**:526
  butterfly and moth **4**:*504, 505*
  honeybee **4**:537
  housefly **4**:*549*

Q

quillwort **2**:284

R

regeneration **7**:988
  starfish and **9**:1228
reproductive system
  7:984–1005
reproductive structures
  in animals
  7:*994–999*
reproductive structures in
  plants **7**:*1000–1003*
timing of **7**:997
zygote development
  7:*1004–1005*

S

scrotum **7**:994
scyphistoma **5**:637, 638,
  646, 647; **7**:*987*
seeds, plant **2**:276–277;
  7:1003
  dispersal **6**:740, 753;
  7:1003, **8**:1058,
  1148–1149
  grass **6**:752–*753*
  orchid **6**:778
  useful **6**:752
semen **7**:995
seminal receptacle **3**:341;
  9:*1264*
seminiferous tubules
  7:*994*
Sertoli cells **7**:*994*, 998
sex
  changing
    bivalves **3**:355
    fish **4**:459; **7**:*998*
  determination
    7:991–992
  temperature and
    9:1206–1207
sexual reproduction
  7:*985*, 990–993
  hermaphroditism
    7:*998*
sexual selection **10**:1379
sori, fern **2**:284, *285*

sperm (spermatozoa)
  1:67, *87*; **7**:*990,
  993, 994,* 995–996,
  1004
  competition **3**:395
spermathecae **4**:504
  of ants **1**:34
  of earthworms **2**:237,
    243
  of houseflies **4**:549
  of newts **6**:811
  of weevils **10**:1357
spermatia **8**:1136
spermatids **7**:993
spermatocytes **7**:993
spermatogenesis **7**:*994*
spermatogonia
  7:992–993, 994
spermatophores **3**:336,
  341; **4**:504; **5**:691,
  699; **6**:*810*;
  8:*1082*
spermatozoids **8**:1137
sporangia
  fern **2**:*283,* 284, *285*
  fungal **6**:779
sporangiospores **6**:777
spores
  bacterial **1**:*53*
  fern, dispersal **2**:284
  fungal **6**:775, 777,
    778–779
  seaweed **8**:1134, 1135
sporocytes **2**:*283,* 284
sporophyte **7**:993, 1000,
  fern **2**:277; **7**:*993*
  seaweed **8**:*1134*
stamens **1**:43; **6**:751–752;
  7:*1001*
stigmata **1**:*43*; **7**:*927*
strobilae **7**:986–987
styles (plant) **1**:*43*; **6**:752

T

testes
  human **4**:*572,* 573;
  7:*994*

internal and external
  3:330; **8**:1047
testes-determining factor
  (TDF) **7**:992
tetraspores **8**:*1134, 1135*
theca cells **7**:995
trochophores **2**:243

U

urodeum **10**:1416
uterus **4**:*572*
  bicornuate **3**:372;
    6:742; **7**:945;
    8:*1046,* 1124;
    10:*1378,* 1403
uteri of kangaroos **5**:*661*;
  7:*996*

V

vagina, of kangaroos
  5:*661*; **7**:*996*
vasa deferentia
  4:572–573; **7**:995
viviparity, in tsetse flies
  4:549

Z

zoospores **8**:1134,
  1136–1137
zygote **7**:990–991

Respiratory system

A

air sacs (parabronchi)
  of birds **3**:307; **8**:1020,
    1028
    albatrosses **1**:*17*
    eagles **2**:231; **8**:*1019*
    hummingbirds **5**:588
    ostriches **6**:841
    woodpeckers
      10:1414
alveoli **4**:*567*; **5**:677;
  7:911; **8**:*1012,* 1020

anal sacs, of turtles
  9:1204, 1205

B

bends, the **2**:203; **8**:1121
Bernoulli effect **8**:1026
blowholes **2**:195, *196,*
  202, *203*; **3**:380,
  386, 391; **8**:*1014*
book lungs **8**:1015,
  1022, 1080;
  9:1257, *1261*
breathing air using lungs
  8:1013, *1019–1020*
breathing underwater
  8:1013, *1016–1018*
breathing, voluntary and
  involuntary **4**:520;
  8:1020
bronchi **4**:566–*567*;
  5:677; **7**:911;
  8:1020; **10**:1374
bronchioles **4**:566–*567*;
  5:677; **8**:1020

C

chlorocruorin **8**:*1027*

D

decompression sickness
  see bends, the
diaphragm **3**:329; **4**:567;
  5:677; **7**:911,
  958–959; **8**:*1028*;
  9:1219; **10**:1432

G

gas exchange **8**:*1012,*
  1013–1015
gills **8**:1016–1017,
  1029–1030
  of crabs **1**:138
  dermal **9**:1227
  of fish **2**:176
    coelacanths **1**:*129*
    gulper eels **3**:*443*
    hagfish **4**:*455*

hammerhead sharks
4:472–473
sea horses 8:1105
sharks 8:1027
trout 9:1289
tuna and billfish
8:1071
of giant clams 3:345,
347, 354; 8:1016
of insects 8:1017
of octopuses 6:821
ventilation
8:1026–1027
glottis 8:1019

H

hemocyanin 3:338;
8:1015, 1027

I

infundibulum 5:591
intercostal muscles
3:329; 8:1028

L

lactate dehydrogenase
7:875, 878
lactic acid 5:678; 7:875,
878; 8:1012, 1121
lungs 8:1030
of birds 8:1020
eagles 2:231
of elephants 2:259
evolution 1:129
of snakes 3:407

M

maximizing efficiency
8:1013, 1029–1031

P

palpal organs 9:1264,
1265
papulae, of starfish
9:1233
pigments 8:1013, 1015,
1023–1025

R

respiration
aerobic 4:472; 5:677,
678; 7:878; 8:1012
anaerobic 5:678;
8:1012
by burrowing sea
anemones 8:1091
by turtles 9:1204
buccopharyngeal 6:809
cellular 4:566; 7:910
cutaneous (through
the skin) 6:804;
8:1013–1014,
1016
respiratory system
8:1012–1031

S

shell gland 7:881
spiracles 8:1021
of dragonflies 2:213,
217
of fish
bottom-living
sharks 4:473
rays 9:1240, 1248
of lice 5:696
of moths 4:503
of myriapods and
centipedes 3:333,
336, 338
of scorpions 8:1078
of weevils 10:1353,
1355–1356
swim bladder 6:765
in coelacanths 1:129
in sailfish 8:1072
in sea horses 8:1105
in trout 9:1286

T

taenidia 8:1021–1022
trachea (windpipe)
4:566; 5:677;
8:1020; 10:1374

tracheae
of invertebrates 1:107;
4:503; 8:1012,
1015, 1021–1022
ants 1:30
centipedes 3:336,
338
dragonflies 2:213,
214, 217
honeybees 4:535
houseflies 4:547
ventilation 8:1026
weevils 10:1353,
1356
tracheal system 8:1013,
1021–1022
see also tracheae
tracheoles 8:1022;
10:1351, 1353,
1356
trophonemata 9:1251
turbinates, nasal 8:1031

V

velar folds 4:455
ventilation 8:1026–1028
see also respiration

Skeletal system

A

acetabula 3:300
antennules 1:135
antlers 7:965, 966, 969,
971–972, 974, 975,
982; 9:1171
females with 7:982
apodemes 2:214;
9:1258
atlas bone (a vertebra)
4:557; 6:731;
9:1173; 10:1391
axis bone (a vertebra)
4:557; 9:1173

B

beak, bird 9:1170
bone
cells 9:1168, 1184
density 9:1168
of humans,
development of
long bones 9:1169
marrow see marrow,
bone
of sirenians 5:707
spongy
in elephants 2:250,
251
in whales 2:194

C

cannon bone 3:362;
7:970
carapace 1:134–135
of lobsters 5:686
of spiders 9:1254
of tortoises 9:1268,
1269, 1272
of turtles 9:1197–1198,
1200
carpals 9:1174
carpometacarpus 7:873
carpus 10:1425
cartilage 9:1168–1169
hyaline 9:1169
in sharks 4:464, 465;
9:1169, 1242
in skates and rays
9:1242
centrum 9:1172
cephalothorax see
prosoma
chelae 1:135, 139; 5:687;
8:1077
chelipeds 5:687
chitin 1:135; 2:210–211;
4:496, 543; 9:1257
clavicle (collarbone)
9:1174
of chimpanzees 1:93

clavicle *(continued)*
  of monotremes **7**:889
  of pumas **7**:937
coccyx **1**:93; **4**:558;
    9:1173
columella **6**:*838*; **9**:1170
coracoid bones **5**:585;
    7:*873*, 888–889;
    9:1173, 1174,
    1200
cranium **7**:938
cuticle **9**:1258
  of dragonflies **2**:210
  of lice **5**:695
  of lobsters **5**:687;
    9:*1167*
  of spiders **9**:1257

D

deltoid processes **1**:96
digitigrade stance **3**:418
deer **7**:970
  elephants **2**:253
  Eurasian otters **6**:852
  hares **4**:*483*
  hyenas **5**:596
  lions **5**:670
  monkeys and apes
    6:733
  wolves **10**:1390

E

elytra **2**:211; **4**:542
  of beetles
    10:1346–1347,
    1356
endocuticle **2**:210, 213;
    9:1258
endosterna **9**:1258
endosternite **9**:1258
epicuticle
  dragonfly **2**:211, 213
  lobster **5**:687
epipubic bones **5**:654,
    655; **7**:889
  evolution **9**:1157–1160

exocuticle **2**:210, 213;
    9:1258
exoskeleton
  of arthropods **9**:1165
    centipedes **3**:334
    dragonflies
      2:210–211
    honeybees
      4:529–530
    houseflies
      4:542–543
    lice **5**:695
    scorpions
      8:1076–1077
    spiders **9**:1257
  of crustaceans **9**:1163
    crabs **1**:*134*–135
    lobsters **5**:*686*

F

femur **9**:*1174*
  seal **8**:1116
fibula **9**:*1174*
  deer **7**:*970*
fontanelle **9**:1171
foramen magnum **4**:556
furcula (wishbone) **1**:14;
    7:*873*, 874

G

glenoid fossa **10**:1411

H

hallux, opposable **6**:729
halteres **2**:211; **4**:542
Haversian canals **9**:1168
hemal arches **9**:1172
hooves
  giraffe **3**:*363*
  hippopotamus **4**:*510*,
    513–514
  zebra **10**:*1421*–1422
horns **7**:*972*; **9**:1171;
    10:1364–*1365*,
    1367, 1369
  females with **10**:1379

humerus
  albatross **1**:*14*
  eagle **2**:226
  human **4**:558
  seal **8**:1116
  whale **2**:196
hydrostatic/hydraulic
    skeleton **3**:323;
    9:*1162*–1164
  of caterpillars **4**:501
  of earthworms
    2:239; **9**:*1163*–1164
  of jellyfish **5**:642;
    9:*1162*
  of sea anemones
    6:*755*
hyoid apparatus
  of birds **10**:*1411*
    hummingbirds **5**:589
  of elephants **2**:253
  of small cats **7**:*935*
  and the tongue **3**:319
hyoid arch **9**:1171

I

incus **4**:555; **9**:1170
infraorbital foramen
    7:952
interclavicle **7**:888–889
intervertebral disks
    9:1172
intracranial joints **1**:*126*
invertebrate
    9:*1162–1167*

K

keel **1**:14; **7**:873
keratin **10**:1366

L

limbs, pentadactyl
    9:*1174*–1175

M

malleus **4**:555; **9**:1170
mandibular arch **9**:1171

marrow
  bone **1**:*108*;
    4:555–556; **5**:618;
    9:1168
  stem cells **7**:959
masseteric tubercle
    9:1216
mesocuticle **9**:1258
mesohyl **9**:1162
mesothorax **10**:1348
metacarpals **3**:298;
    10:1390, 1426
metatarsals **10**:1390,
    1424, 1426
metathorax **10**:1348

N

neural plates **9**:1272
neural spines **9**:*1172*
notochord **4**:476;
    9:1158, 1172
  of hagfish **4**:*448, 452*;
    9:*1159*

O

occipital condyles **9**:1173
olecranon process **8**:1039
ossicles
  ear **4**:555, *565*
  scleral **9**:1199
  of a starfish skeleton
    9:1228
ossicones **3**:359–360, *363*
osteoporosis **9**:1168

P

parietal bone, in
    chameleons **5**:628,
    629
pectoral girdle **9**:1174
  of otters **6**:852
  of rays **9**:*1242*, 1243
pedicels **7**:*970*, 971, 972;
    9:1257
pelvis (pelvic girdle)
    9:1174

frog **9**:1159–1160
otter **6**:852
vestigial **9**:*1175*
zebra **10**:1426
pentadactyl limbs **5**:629
periosteum **7**:971–972
phalanges
  bat **3**:298
  chameleon **5**:*629*
  giant anteater **3**:321
  giraffe **3**:363
  human **4**:558
  seal **8**:1116
  whale **2**:196
plant **9**:*1161*
plantigrade stance
  bears **3**:*418*
  hares **4**:483
  humans **4**:553
  monkeys **6**:733
  otters **6**:852
  rats **7**:953
  squirrels **9**:1214
pleopods **1**:135
processus spinalis **1**:93;
    6:732
procuticle, dragonfly
    2:210
prosoma (cephalothorax)
    1:134; **8**:1076;
    9:1252–1253,
    1254
prothorax **10**:1348
pterostigma **2**:211–212
pterygoid bones **3**:319
pygostyle **6**:838; **7**:873;
    10:1410, 1411

R
radius
  albatross **1**:*14*
  eagle **2**:226
  seal **8**:1116
resilin **4**:543

S
sacrum **4**:558; **6**:732;
    7:935; **9**:1173;
    10:1391
sagittal crest **1**:92; **4**:514
scapulae **9**:1174
  of chimpanzees **1**:93
  of giant anteaters **3**:321
  of giraffes **3**:363
  of montremes **7**:889
  of zebras **10**:1426
semi-digitigrade posture
    4:513
sesamoids, of giraffes
    3:363
shell, giant clam
    3:*344*–345, 349
shoulder girdle, of turtles
    9:1200
skeletal system
    9:**1156–1175**
  skull **9**:*1170–1171*
  see also bone; cartilage
  see also exoskeleton
skull, vertebrate
    9:*1170–1171*
snout, dental **6**:733
squamosal bone, in
    chameleons **5**:628,
    629
stapes **4**:555; **9**:1170
sternum
  of birds **9**:1173
    albatross **1**:14
    eagle **2**:226
    hummingbird **5**:*585*
    ostrich **6**:*838*, 839
    woodpeckers
      10:1410
  human **4**:558
  of spiders **9**:1254
supraorbital processes
    4:482
syndactyl animals **5**:655
synsacrum **7**:873; **9**:1173;
    10:1411

T
tarsometatarsus **6**:839
tarsus **10**:*1425*
tendrils, plant **9**:1161
tergites **1**:134; **6**:*762*
thumbs, opposable **1**:*97*;
    4:553; **6**:729, *733*
tibia **9**:*1174*
tibiofibula **1**:59
transverse processes
    9:*1172*
tube feet **9**:1164, *1227*,
    1229, *1230*, 1232

U
ulna
  albatross **1**:*14*
  deer **7**:*970*
  eagle **2**:226
  seal **8**:1116
ulnar sesamoid **3**:298
urostyle **9**:1160, 1173

V
vertebrae **9**:1172–*1173*
  of anteaters **3**:*320*
  caudal **5**:707; **7**:935;
    9:1173, 1214
  cervical **1**:58; **4**:557;
    6:731; **7**:934;
    9:1173; **10**:1368,
    1410, 1425
  dorsal **5**:654
  of giraffes **3**:363;
    9:*1173*
  of gray whales
    3:381–*382*
  of hippopotamuses
    4:512
  of humans **4**:557–558
  lumbar **4**:558; **5**:654;
    6:732; **7**:934–935;
    8:1038, *1116*;
    10:1410–1411
  postsacral **1**:59
  sacral **1**:59

of sharks and rays
    9:1172
of snakes **3**:*402*
thoracic **4**:557–558;
    6:731, 732; **7**:934;
    8:1038; **9**:*1172*,
    1173; **10**:1410,
    1425
vertebrate **9**:*1168–1175*

W
water vascular system
    9:*1229*, 1233
wings **3**:*299*
  bat **3**:294–*295*,
    299–300, 301–302
  dragonfly **2**:211–212
  honeybee **4**:529
  housefly **4**:544
  insect **2**:211

X
xenarthrales **9**:1185

Z
zygantra **3**:402
zygapophyses **1**:58;
    3:402; **8**:1116
zygomatic arches **7**:952;
    9:*1216*

Index of geographical place-names

A

Africa
  aardwolves **5**:593
  antelope **10**:*1376*
  chameleons **5**:624
  hyenas **5**:593
  lions **5**:664
  manatees **5**:700
  ostriches **6**:835
  pangolins **9**:*1181*
  porcupines **7**:900, 901, *902, 907*
  scaly-tailed flying squirrels **9**:1213
  scorpions **8**:1075
Alaska, wolves **10**:1385
Arctic Ocean, gray whales **3**:375
Arizona, cacti **8**:*1051*
Asia
  ground squirrels **9**:*1212*
  pangolins **9**:*1181*
  pikas **4**:480
  porcupines **7**:900, 901
  tigers **5**:668
  wolverine **10**:*1330*, 1331
Atlantic, stingrays **9**:*1239*
Australia
  crown-of-thorns starfish **9**:1235
  dingos **7**:885
  echidnas **7**:883
  giant sea anemones **8**:1086
  kangaroos **5**:648
  monotremes and marsupials **7**:885
  pitcher plants **10**:*1304*
  sea anemones **8**:*1091*

B

Borneo
  pitcher plants **10**:*1303*
  porcupines **7**:900

Brazil, sloths **9**:*1183*, 1190, 1193

C

California
  cacti **8**:*1051*
  chameleons **5**:624
Canada, wolves **10**:1385
Caribbean islands, manatees **5**:700
Caribbean Sea, jellyfish **5**:645
Central America
  manatees **5**:700
  sloths **9**:*1183*
Colombia, porcupines **7**:903
Congo, Democratic Republic of, mystery chimpanzee **1**:91

E

Egypt, wolves **10**:1385
Europe
  ground squirrels **9**:*1212*
  wolverine **10**:*1330*

F

Florida, waters off, jellyfish **5**:645

G

Galápagos islands, giant tortoises **9**:1270
Great Barrier Reef, sea anemones **8**:*1091*
Guatemala, Montezuma pine **8**:1140

H

Hawaiian islands, chameleons **5**:624

I

India
  chameleons **5**:624
  hyenas **5**:593
Indian Ocean, sea horses **8**:*1098*
Ireland, weasels **10**:1340

K

Kenya, chameleons **5**:624

M

Malay peninsula porcupines **7**:900
Mexico
  cacti **8**:*1050, 1051*
  Montezuma pine **8**:1140
  onza **7**:945
  tortoises **9**:*1271*
  wolves **10**:1385
  woodpeckers **10**:*1405*
Middle East
  chameleons **5**:624
  hyenas **5**:593

N

New Zealand, kiwis **6**:834
Nicaragua, sloths **9**:*1183*
North America
  grizzly bears **3**:411
  ground squirrels **9**:*1212*
  pikas **4**:480
  wolverine **10**:*1330*
North Carolina, Venus flytrap **10**:1301

P

Pacific Ocean
  gray whales **3**:375
  sea horses **8**:*1098*
Papua New Guinea, echidnas **7**:883
Peru, weevils **10**:*1345*

R

Russia, sables **10**:1329

S

Seychelles, giant tortoises **9**:*1267*
South America
  anacondas **3**:397
  armadillos **9**:*1181*
  hummingbirds **5**:580
  manatees **5**:700
  maned wolves **10**:*1385*
  piculets **10**:1405
  porcupines **7**:900, 902, *903, 912, 913*
  tinamous **6**:835
South Carolina, Venus flytrap **10**:1301
Spain, weasels **10**:1340
Sri Lanka, chameleons **5**:624
Sumatra, porcupines **7**:900

T

Tanzania, chameleons **5**:624

U

United States
  antelope jackrabbits **4**:479
  armadillos **9**:*1181*
  flying squirrels **9**:1213
  manatees **5**:700
  tortoises **9**:*1271*
  turtles **9**:1196
  woodpeckers **10**:*1405*

Index of microbiology, cell biology, and genetics

A

acontia **8**:*1093*

adenosine diphosphate *see* ADP

adenosine triphosphate *see* ATP

ADP (adenosine diphosphate) **6**:759, *760*

algae blue-green *see* cyanobacteria

alleles **1**:76–79

alveolates **2**:*164*

amoeba **1:20–25**; **2**:*164*, 166; **8**:*1126*

  external anatomy **1**:*21, 22*

  internal anatomy **1**:21, *23*

  reproductive system **1**:21, *24–25*

Anabaena **1**:*49*

anemia, sickle-cell **1**:*78*, 79

apical complex **1**:21

apoptosis **1**:87

ATP (adenosine triphosphate) **1**:72, 73; **6**:741

  powering muscles **6**:*759, 760, 761*

autosomes **7**:990

auxospores **2**:171

avian flu virus *see* bird flu virus

axopods **2**:167

B

Bacillus **1**:53

bacteria **1:46–53**

  aerobic and anaerobic **1**:52

  anatomy

    external **1**:47, *48–49*

    internal **1**:47, *50–51*

  autotrophic **1**:52

cellulose-digesting **2**:182

digestive and excretory systems **1**:47, *52*

enteric **1**:47

Gram-positive and Gram-negative **1**:*48–49*

heterotrophic **1**:52

immune system and **5**:606–614

lipopolysaccharide (LPS) coat **5**:610

nitrogen-fixing **1**:47; **2**:172

pathogenic **5**:606, 610

reproductive system **1**:47, *53*

spores **1**:*53*

bacteriophage **10**:*1320, 1321, 1322*

bird flu virus **5**:*614*; **10**:*1319*

C

Candida albicans **5**:*607*

cell biology and genetics **1:64–87**

  beginnings of life **1**:65, 69

  cell constituents **1**:64–65

  cell cycle and cell division **1**:*84–87*

  cell external anatomy **1**:68

  cell internal anatomy **1**:*69–74*

  see also organelles

  cells and their functions **1**:*66–67*

    genes and inheritance **1**:*76–79*

    movement and support **1**:75

    nucleic acids **1**:*80–83*

see also DNA; RNA

cell membrane (cytoplasmic membrane; plasma membrane) **1**:*68*

  bacterial **1**:50, *51, 52*

cell wall **1**:68

  of ciliates **2**:167

  of diatoms (frustule; theca) **2**:166, *167*

  of dinoflagellates **2**:167

  of heliozoans **2**:167

  of radiolarians **2**:167

centromeres **1**:*76*

chlorenchyma **8**:*1055*

chloride cells **9**:1290

chlorophyll **1**:42, 74

chloroplasts **1**:42, *74*; **6**:*750*, 829–830; **7**:923, **8**:*924*; *1145*

  cactus **8**:1055

  diatom **1**:*74*; **2**:169

  sea lettuce **8**:1132

chlorosomes **1**:51

chondroblasts **9**:*1169*

chondrocytes **9**:1169

chromatin **1**:79

chromatophores **5**:*631*; **6**:*789*, 816

chromocytes **9**:1283–1284

chromosomes **1**:*76*; **4**:493; **7**:990

  autosomes **7**:990

  bacterial **1**:50–51

  of fruit flies **4**:543

  and polyploidy **7**:927

  replication **1**:*84*

  sex (X and Y) **1**:*76*, 78; **7**:990, 991

chrysolaminarin **2**:169

cilia **1**:75

ciliates **2**:*164*, 165; **8**:*1126*

  cell wall **2**:167

cnidae **8**:*1092*–1093

cnidocytes **8**:1092

codons **1**:81

collenchyma cells **1**:67; **7**:924

conjugation **1**:24, *25*, 53

crassulacean acid metabolism (CAM) **6**:830; **8**:*1030, 1057*

cyanobacteria (blue-green algae) **1**:*46*, 47, 48, *49*; **8**:1015

  on sloths' fur **9**:*1178, 1182*, 1183

cyclic adenosine monophosphate (cAMP) **2**:265

cyclins **1**:85

cystic fibrosis **1**:79

cytoplasm **1**:73

cytoplasmic membrane *see* cell membrane

cytoskeleton **1**:*73*, 75

  diatom **2**:169–170

cytosol **1**:50

D

deoxyribonucleic acid *see* DNA

desmosomes **6**:766

diatom **1**:*74*; **2:164–171**

  centric **2**:*164*, 165, 168

  colonial **2**:*170*

  and diatomaceus earths **2**:166

  differ from bacteria **2**:168

  external anatomy **2**:165, *166–167*

  fossils **2**:169

  internal anatomy **2**:165, *168–170*; **9**:*1157*

  pennate **2**:*164*, 165

    gliding movement **2**:168

  reproductive systems **2**:165, *171*; **7**:*987*

diatom *(continued)*
  shapes **2:***167*
dinoflagellates **2:***164*, 165;
  8:*1126*
  cell wall **2:**167
diploid cells **1:**76, 87;
  7:985, 990
DNA **1:**65, *79*, *80–81*;
  7:*984*
  bacterial **1:**50–*51*, 53
  kinetoplastid **1:**24
  mitochondrial
    (mDNA) **1:**83;
    8:1146
  mutation **1:**85
  replication **1:***84–87*
  vaccines **1:**83
  see also plasmids
dwarfism **1:**79

E

E. coli see Escherichia coli
ectoplasm, of amoebas
  1:23
electron transport chain
  1:*72*
endodermal cells **8:**1145
endoplasm, of amoebas
  1:23
endoplasmic reticulum
  1:*69*, 70–71
  rough **1:**70–71
    of diatoms **2:**169
  smooth **1:**71
endosymbiont theory
  1:69
Entamoeba histolytica **1:**21
epicingulum **2:***167*
epithelial cells **1:**67
Escherichia coli (E. coli)
  1:*47*, *48*, *50*, *51*, *80*
  viruses attacking
    10:*1320*
eukaryotes **1:**69; **2:**170
extracellular matrix
  (ECM) **1:**68

F

fimbriae (pili) **1:***47*, 49
flagella **1:**75
  of bacteria **1:**49
  of diatoms' sex cells
    2:*170*, 171
flagellates **2:***164*, 165
foraminiferans **1:**22;
  2:*164*, 165
  cell coverings **2:**167
frustule *see* theca

G

gap junctions **6:**766
gas vesicles, bacterial
  1:51
genes **7:***984*
  alleles **1:**76–79
  and altruism **10:**1401
  Hox **7:**1005
  recessive and dominant
    1:78
  S-genes **1:**45
  "wild-type" **1:**78
genotypes **1:**77
Giardia lamblia **1:**20
glycoproteins **10:**1318
goblet cells **2:***181*
Golgi apparatus **1:***71*
  of diatoms **2:**169
grana **6:***750*
guard cells **1:**42

H

haploid cells **1:**76, *87*;
  7:985, 990
heliozoans **1:**21; **2:***164*,
  167
herpesvirus **10:***1322*
heterocysts **1:**49
HIV (human
  immunodeficiency
  virus) **5:***616*, *620*;
  10:*1315*, *1323*
Human Genome Project
  1:77

human immuno-
  deficiency
  virus *see* HIV
hypocingulum **2:***167*
hypovalve, of diatoms
  2:166

I

inclusion granules,
  bacterial **1:**51
influenza virus **10:***1317*
iridiocytes **9:**1284
iridocytes **6:**816

K

kinases, cyclin-dependent
  1:85
kinetoplast **1:**21

L

life, beginnings of **1:**64,
  69
lysosomes **1:**73

M

meiosis **1:***67*, *84*, 87
  and diatom
    reproduction **2:**171
melanophores **5:***631*;
  6:*789*
Mendel, Gregor **1:**77–78
mesogleal cells **8:**1089
mesophyll cells **7:***924*;
  8:*1145*
microorganisms
  excretion by **2:**187
  plant- and cellulose-
    digesting **2:**181, 182
  in elephants **2:**261
microtubules **1:***75*
  diatom **2:**169–170
mitochondria **1:***69*,
  72–73; **5:**586;
  6:*767*;
  9:1188, 1290
  of diatoms **2:**169

DNA **1:**83; **8:**1146
  in muscle cells **7:**875;
    8:1067
mitosis **1:**25, 84–87;
  7:991
mutations **1:**85, 86–87
  in fruit flies **6:**793
mycobacteria **5:**606
mycoplasmas **1:**48

N

natural selection **1:**86
nephrocytes **9:**1263
neuromasts **9:**1247
nucleus **1:**70

O

organelles **1:***64*, *69*,
  70–74
  of diatoms **2:**168–169
  evolution from
    prokaryotes **1:**69;
    2:170
osteoblasts **9:***1168*
osteoclasts **9:**1168

P

paramecium **1:**21, 22, *25*,
  75; **2:**167
parenchyma cells **1:**67;
  6:829, *830*, 831;
  7:924, 925;
  8:*1055*–1056, 1144;
  10:1306
parenthosome **6:**776
peptidoglycan **1:**48–49
peroxisomes **1:**73
phagocytosis **1:**23, *66*
phenotypes **1:**77
  wild-type **1:**78
photocytes **10:**1349
photosynthesis **1:**74;
  8:1013; **10:**1307
  C3 and C4 **6:**750;
    8:1031
picornavirus **10:***1321*

pili *see* fimbriae
pinocytosis **1**:*66*
plasma membrane *see* cell
   membrane
plasmids **1**:*51*, 53, 77, *81*
plasmodia **8**:*1126*
Plasmodium falciparum
   5:*607*
polyploid cells **1**:76;
   7:*927*
poxvirus **10**:1322
prion **10**:*1318*
prokaryotes **1**:20, *46*, 50,
   64; **2**:170
   mitosis in **1**:*85*
protists **1**:20; **2**:164–165
protozoans **1**:20
   amoeboid **1**:*20*, 21
   ciliated **1**:*20*, 21
   flagellated **1**:*20*
   parasitic **1**:*20*, 21
pseudopodia **1**:*21*, 22
pyruvate (pyruvic acid)
   1:*72*, 73

R

radiolarians **1**:21; **2**:165
   cell wall **2**:167
raphe, of diatoms **2**:166
retroviruses **10**:*1323*
ribonucleic acid *see*
   RNA
ribosomes **1**:*69*; **7**:*984*
   bacterial **1**:50, *70*
Rickettsia **5**:*607*
RNA **1**:65, 80, 81
   and the first forms of
      life **7**:*984*
   messenger (mRNA)
      1:70, 80, 81–82, *83*;
      7:*984*; **10**:1320,
      1321
   ribosomal (rRNA)
      1:50, 82
   transfer (tRNA)
      1:*82–83*

rods and cones **6**:*786*,
   791; **8**:1069;
   9:1274, 1287;
   10:1396

S

Salmonella typhosa **5**:*607*
sarcodines **2**:*164*
sclerenchyma cells **1**:67
S-gene **1**:45
siphonoglyphs **8**:1088
sleeping sickness **1**:21
slime mold **8**:*1126*
sporozoans
   (apicomplexans)
   1:21, 24
staphylococci **5**:608
stem cells **7**:*959*
stramenopiles **2**:*164*
Streptococcus pneumoniae
   5:*607*

T

T–bacteriophage virus
   10:*1320*, *1321*
tetraploid plangts **7**:*927*
theca (frustule), of
   diatoms **2**:166, *167*
thylakoids **6**:*750*
tobacco mosaic virus
   10:*1316*, 1317
tracheids **7**:*924*; **8**:1144;
   9:1161
transcription and
   translation
   1:*81–83*; **10**:1320
trumpet hyphae cells
   8:1133
Trypanosoma brucei **1**:21
typanosomes **1**:21

V

vaccines, DNA **1**:83
vacuoles
   of amoebas **1**:23
   plant **1**:74

virions **10**:1314, *1317*
viroids **10**:1317
virus **5**:606;
   10:**1314–1323**
   antigenic drift and shift
      5:621
   classification
      10:1314–1315
   DNA viruses **10**:1314,
      1321, *1322*
   immune system and
      5:609, *613–614*
   retroviruses **10**:*1323*
   RNA viruses **10**:1314,
      1320, *1321*
   structure **10**:1315,
      1316–1318
   within cells
      10:*1319–1322*

Z

zooxanthellae
   in giant clams **3**:342,
      343, 352
   in sea anemones
      8:1092

Index of scientific names

A

Acanthaster planci **9**:*1224*
Acanthostega **9**:1160
Acarina **8**:*1074*–1075; **9**:*1252*
Accipitridae **2**:*220*
Acrosiphonia coalita **8**:1131
Actinaria **8**:*1084*, 1085
Actinopoda **1**:21, 22
Actinopterygii **3**:*436*, 437; **8**:*1060*, 1061, *1096*, 1097; **9**:*1174*, *1280*, 1281
Adenoviridae **10**:1314
Adephaga **10**:*1344*, 1345, 1357
Adetomyrma **1**:*26*, 33
Aepyornis maximus **6**:837
Aeshnidae **2**:*208*
Agaricaceae **6**:*770*, 771
Agaricales **6**:*770*, 771
Agaricus
  bisporus **6**:*770*, 771
  campestris **6**:*770*
Agnatha **3**:*436*; **4**:*446*, *460*; **8**:*1060*, *1096*; **9**:*1238*, *1280*
Ailuropoda melanoleuca **3**:*410*
Aizoaceae **8**:1049
Alaria esculenta **8**:*1130*
Albuliformes **3**:*436*
Aldrovanda **10**:1301
Alismatidae **6**:825
Alligatoridae **2**:148
Alveolata **2**:*164*
Amanita
  muscaria **6**:771
  phalloides **6**:771
  virosa **6**:771
Amanitaceae **6**:*770*
Amblycera **5**:*692*–693
Ambystoma **6**:801, 811
Ambystomatidae **6**:*800*, 801

Ammonoidea **6**:815
Amoeba **2**:*164*
Amoeba proteus **1**:*20*, 22
Amoebidae **1**:*20*
Amphibia **1**:*54*; **6**:*800*
Amphisloaemia **5**:*624*
Amphiumidae **6**:801
Anax imperator **2**:*208*
Androctonus **8**:1081
Angiospermae **1**:*36*; **6**:*824*; **7**:*916*, 917; **8**:*1048*, *1138*
Anguilliformes **3**:*436*
Anisoptera **2**:*208*
Anisozygoptera **2**:*208*
Anomalodesmata **3**:*342*
Anomaluridae **9**:*1208*
Anomura **1**:*132*, 133, 135
Anoplura **5**:*692*, 693, 694
Anthophoridae **4**:*526*
Anthozoa **5**:*636*; **8**:*1084*
Anthribidae **10**:*1344*
Antilocapridae **10**:1371
Antilopinae **10**:*1360*
Antipatharia **8**:*1084*
Anura **1**:*54*, 60; **6**:*800*
Apatelododae **4**:*494*
Aphelocheirus **8**:1018
Aphonopelma **9**:1265
Aphyllophorales **6**:771
Apiales **1**:*36*, 37
Apicomplexa **1**:*20*, 21
Apidae **1**:*26*; **4**:*526*, 527
Apinae **4**:*526*
Apis **4**:527
  andreniformis **4**:*526*
  cerana **4**:*526*
  dorsata **4**:*526*
  florea **4**:*526*
  koschevnikavi **4**:*526*
  laboriosa **4**:*526*
  mellifera **4**:*526*
Aplacophora **3**:*342*; **6**:*812*
Aplysia **6**:819

Apocrita **1**:*26*, 27; **4**:*526*, 527
Apodiformes **5**:*580*
Apodimorphae **5**:*580*
Apoidea **4**:*526*, 527
Apostasioideae **6**:*824*, 825
Aptenodytes **7**:*868*
Aquificae **1**:*46*, 47
Arachnida **8**:*1074*; **9**:*1252*, 1253
Araneae **8**:*1074*; **9**:*1252*, 1253
Araneomorphae **9**:*1252*, 1253
Araucaria **8**:1139
Araucariaceae **8**:1139
Archaea **1**:*46*
Archaeobatrachia **1**:*54*, 55
Archaeopteryx **9**:*1160*
Archilochus colubris **5**:*580*
Archostemata **10**:*1344*, 1345
Arctodus **3**:415
Arecidae **6**:825
Arenaviridae **10**:1314
Arenavirus **10**:1314
Arethuseae **6**:825
Ariocarpus **8**:1052, 1055
Armillarea bulbosa **6**:772
Arthrobotrys **6**:774
Arthropoda **1**:*26*, *132*; **2**:*208*; **3**:*332*; **4**:*494*, *526*, *538*; **5**:*684*, *692*; **8**:*1074*; **9**:*1252*
Artiodactyla **3**:*356*; **4**:*506*; **7**:*964*; **8**:*1032*; **10**:*1360*, 1361, *1418*
Arvicolinae **7**:*946*
Aschiza **4**:*538*
Ascomycetes **6**:776, 779
Ascomycota **6**:*770*, 771, 775

Asplenium bulbiferum **2**:285
Asterias forbesi **9**:*1237*
Asteridae **7**:*916*
Asteroidea **9**:*1224*
Athyrium **2**:276
Atropa belladonna **7**:*916*
Atta **1**:*26*
Attelabidae **10**:*1344*
Aurelia aurita **5**:*636*
Auricularia **6**:771
Australopithecus **4**:551
Aves **1**:*10*; **2**:*220*; **5**:*580*; **6**:*834*; **7**:*868*, 869; **10**:*1404*
Azolla **2**:280

B

Babyrousinae **4**:*506*
Bacillariophyta **2**:*164*
Balaenoptera
  musculus **3**:*374*
  omurai **3**:*374*
  physalus **3**:*374*
Balaenopteridae **3**:*374*
Bangiophycidae **8**:1127
Basidiomycetes **6**:776, 779
Basidiomycota **6**:*770*, 771
Basilosaurus **2**:192
Batoidea **9**:*1238*, 1239
Belidae **10**:*1344*
Bivalvia **3**:*342*; **6**:*812*
Blossfeldia lilputana **8**:1050
Boidae **3**:396
Boletaceae **6**:*770*
Bombinae **4**:*526*
Bombycidae **4**:*494*
Bombycoidea **4**:*494*
Bovidae **10**:*1360*, 1361
Bovinae **10**:*1360*, 1361
Bovoidea **3**:*356*; **7**:*964*
Brachiosaurus **9**:1160
Brachycephalidae **1**:*54*
Brachycera **4**:*538*, 539

Brachylagus idahoensis **4**:*476*

Brachyura **1**:*132*, 133; **5**:*684*

Bradypodidae **3**:*312*; **9**:*1176*

Bradypus **9**:1180
  torquatus **9**:*1176*
  tridactylus **9**:*1176*
  variegatus **9**:*1176*

Brentidae **10**:*1344*

Brisingida **9**:*1224*

Brookesiinae **5**:*624*

Bryopsis corticulans **8**:1133

Bufonidae **1**:*54*, 55

Bulbophyllum **6**:832
  globuliforme **6**:828

Bunolagus monticularis **4**:476

C

Cactaceae **8**:*1048*, 1049

Cactoideae **8**:*1048*, 1049, 1052

Calliphoridae **4**:*538*

Callitrichidae **6**:725

Campanulales **7**:*916*

Cancer
  antennaris **1**:*132*
  irroratus **1**:*132*
  jordani **1**:*132*
  magister **1**:*132*
  pagurus **1**:*132*

Cancridae **1**:*132*

Canidae **3**:*410*; **6**:*844*; **10**:*1382*, 1384

Canis **10**:*1382*, 1384
  lupus familiaris **10**:1388
  rufus **10**:1388

Canoidea **5**:*592*, 664; **7**:*928*; **10**:*1324*

Capreolus capreolus **7**:*964*

Caprolagus hispidus **4**:*476*

Carabidae **10**:*1344*

Carcharhinidae **4**:*460*

Carcharhiniformes **4**:*460*, 461

Cardiidae **3**:*342*

Carnegiea gigantea **8**:*1048*

Carnivora **3**:*410*; **5**:*592*, 664, 665; **6**:*844*, 845; **7**:*928*, 929; **8**:*1110*–1111; **10**:*1324*, *1382*, 1383

Carretochelyidae **9**:*1194*, *1266*

Caryophyllaceae **8**:1049

Caryophyllales **8**:*1048*, 1049

Cassiopea **5**:638, 645

Casuariformes **6**:*834*

Catarrhini **1**:*88*; **4**:*550*; **6**:*724*

Caudata **1**:*54*

Caulerpa **8**:1127

Caviidae **7**:*898*

Cebidae **6**:725

Centrales **2**:*164*

Cephalaspidomorphi **4**:*446*, 447

Cephalocereus **8**:1052

Cephalophinae **10**:*1360*, 1361

Cephalopoda **3**:*342*; **6**:*812*

Cephalotus follicularis **10**:*1300*

Ceratodontidae **1**:*122*

Ceratodontiformes **1**:*122*

Ceratotherium simum **8**:*1032*

Cercocebus **6**:*724*

Cercopithecidae **1**:*88*; **4**:*550*; **6**:*724*, 732

Cercopithecus **6**:*724*

Cereus lamprospermus **8**:*1050*

Ceriantharia **8**:*1084*

Cervoidea **7**:*964*

Cervus elaphus **7**:*964*

Cetacea **2**:*188*; **3**:*374*; **4**:*506*

Chamaeleonidae **5**:*624*

Chamaeleoninae **5**:*624*

Cheirostylis **6**:826–827

Chelicerata **9**:*1252*–1253

Chelonia **9**:*1194*, *1266*

Cheloniidae **9**:*1194*, *1266*

Chelydridae **9**:*1194*, *1266*

Chenopodiaceae **8**:1049

Chilopoda **3**:*332*

Chinchillidae **7**:*898*

Chironex **5**:640

Chiroptera **3**:*292*

chiton **3**:*342*

Chlamydiae **1**:*46*, 47

Chloridoideae **6**:*744*, 745

Chlorophyta **2**:*164*; **8**:*1126*, 1127

Choloepus **9**:*1179*–1180
  didactylus **9**:*1176*
  hoffmanni **9**:*1176*

Cholus cinctus **10**:*1351*

Chondodendron tormentosum **6**:761

Chondrichthyes **3**:*436*; **4**:*460*; **8**:*1060*, 1096; **9**:*1238*, *1280*

Chordata **2**:*220*; **8**:*1110*

Chytridiomycota **6**:*770*

Ciliophora **1**:*20*, 21; **2**:*164*; **8**:*1126*

Cingulata **3**:*312*

Cleistes **6**:826

Clitellata **2**:*234*

Cnidaria **5**:*636*; **8**:*1084*, 1085, 1087

Coccinellidae **10**:*1344*, 1345

Codiaeum variegatum **10**:1303

Coelacanthiformes **1**:*122*

Coelodonta **8**:*1036*

Coleoidea **6**:*812*, 813

Coleoptera **4**:*494*, *538*; **5**:*692*; **10**:*1344*, 1345

Colletidae **4**:*526*

Commelinidae **6**:825

Comphotherium **2**:247

Concentricycloidea **9**:*1224*

Coniferophyta **8**:*1138*

Coniophora **6**:*779*

Connochaetes
  gnou **10**:*1360*
  taurinus **10**:*1360*

Copiapoa **8**:*1054*, 1055
  conglomerata **8**:*1050*

Coprinaceae **6**:*770*

Corallimorpharia **8**:*1084*

Coronatae **5**:*636*

Coronaviridae **10**:1314

Coronavirus **10**:1314

Cortaderia selloana **6**:751

Coryanthes **6**:833

Corybas **6**:832

Crataegus **1**:*36*, 37

Crenarchaeota **1**:*46*

Cricetinae **7**:*946*

Crinoidea **9**:*1224*, 1225

Crocodylia **2**:*148*; **5**:*624*; **9**:*1194*, *1266*

Crocodylidae **2**:*148*

Crocodylus **2**:*148*

Crocuta crocuta **5**:*592*

Crossaster papposus **9**:*1224*

Crustacea **1**:*132*–133; **5**:*684*; **8**:*1074*; **9**:*1252*

Cryptobranchidae **6**:801

Cryptodira **9**:*1194*, *1266*, 1273

Ctenophora **5**:*636*–637

Cubozoa **5**:*636*, 638; **8**:*1084*

Cupiennius **9**:1265
Cupressaceae **8**:*1138*, 1139
Curculionidae **10**:*1344*, 1345
Cyathea **2**:280
Cycadophyta **8**:*1138*
Cyclopes spp. **3**:*312*
Cyclorrhapha **4**:*538*, 539
Cydonia **1**:*36*
Cyemetoidei **3**:*436*, 437
Cylindropuntia **8**:*1049*
Cynomys **9**:*1208*
Cyperaceae **6**:*744*
Cyprepedioideae **6**:*824*, 825
Cypripedium **6**:832

D

Dactylaria brochopaga **6**:774
Dama dama **7**:*964*
Dasypodidae **3**:*312*
Decapoda **1**:*132*, 133; **5**:*684*, 685
Decapodiformes **6**:*812*
Delphinidae **2**:*188*
Dendrobium cucumerium **6**:828
Dendrobranchiata **5**:*684*
Dendrolagus **5**:649
Dennstaedtiaceae **2**:*276*
Dermaptera **3**:297
Desmatosuchus **2**:*155*
Dicamptodontidae **6**:801
Dicerorhinus sumatrensis **8**:*1032*
Diceros bicornis **8**:*1032*
Dicksonia **2**:280
Dicotyledonae **1**:*36*; **6**:*744, 824*; **7**:*916*; **8**:*1048*
Dilleniidae **10**:*1300*, 1301

Dinophyta **2**:*164*
Dinornithiformes **6**:*834*
Diomedea **1**:11
  albatrus **1**:*10*
  chlororhynchos **1**:*10*
  chrysostoma **1**:*10*
  epomophora **1**:*10*
  exulans **1**:*10*
  immutabilis **1**:*10*
Diomedeidae **1**:*10*
Dionaea **10**:1301
  muscipula **10**:*1300*
Diplodocus **9**:1160
Diplopoda **3**:*332*
Dipodidae **7**:*946*
Diprotodontia **5**:*648, 655*
Diptera **4**:*538*, 539; **5**:*692*
Discomycetes **6**:*770*
Donacidae **3**:*342*
Drosera **10**:*1300*, 1301
Droseraceae **10**:*1301*
Drosophila melanogaster **4**:*543*
Drosophyllum **10**:*1301*
Dugong dugon **5**:*700*
Dugongidae **5**:*700*

E

Echinocactus platyacanthus **8**:*1050*
Echinodermata **9**:*1224*–1225
Echinoidea **9**:*1224*, 1225
Edentata **9**:1185
Eira **10**:*1324*
Elasmobranchii **9**:*1238*, 1239
Elasmotherium **8**:*1036*
Elephas **2**:*244*
  maximus **2**:*244*
Elopiformes **3**:*436*
Elopomorpha **3**:*436*
Enaliarctos mealsi **8**:*1113*
Endopterygota **1**:*26*

Enhydra lutris **6**:*844*
Enteromorpha **8**:1127, 1131
Enterovirus **10**:1314
Epidendroideae **6**:*824*
Epostoa **8**:1053
Eptatretinae **4**:448
Eptatretus **4**:*446*, 448, 451, 455, 458
Equidae **8**:*1032*; **10**:*1418*
Equus **10**:*1418*
  caballus **10**:1419, 1423
Erethizontidae **7**:*898*
Ericales **10**:1301
Erycinidae **3**:*396*
Eschrichtidae **3**:*374*
Eschrichtius robustus **3**:*374*
Esociformes **9**:*1280*
Eubacteria **1**:*46*
Eubalaena
  australis **3**:*374*
  glacialis **3**:*374*
Eudyptes **7**:*868*
Eudyptula **7**:*868*
Eunectes
  beniensis **3**:*396*
  deschauenseei **3**:*396*
  murinus **3**:*396*
  notaeus **3**:*396*
Eupharyngidae **3**:*436*
Eupholis **10**:*1358*
Eurotamandua **3**:315
Eurotrochilus inexpectatus **5**:585
Euryarchaeota **1**:*46*
Eurypelma **9**:1258, *1261*
Eurypharynx pelecanoides **3**:437, 438
Eutheria **1**:*88*; **2**:*188*, 244; **3**:*374*; **4**:*506*; **5**:*648, 700*; **6**:*724, 844*; **8**:*1032*; **10**:*1418*

F

Fabales **1**:*36*, 37
Falconidae **2**:*220*
Falconiformes **2**:220
Felidae **5**:*592*; **6**:*844*
Feliformia **10**:*1324*
Felis **5**:*664*; **7**:*928*, 929
  concolor **7**:*928*
  lynx **7**:*928*
  pardalis **7**:*928*
  silvestris **7**:*928*
Feloidea **5**:*592, 664*; **7**:*928*; **10**:*1324*
Ferocactus **8**:1052
Filicales **2**:*276*
Firmicutes **1**:47
Fissipedia **3**:*410*; **10**:*1382*
Florideophycidae **8**:1127
Foraminifera **1**:21, 22; **2**:*164*
Forcipulatida **9**:*1224*
Formicidae **1**:*26*
Formicinae **1**:27
Fucus **8**:1131, 1135
  vesiculosus **8**:*1126*
Fungi **6**:770

G

Galaxiidae **9**:1281
Galictis **10**:*1324*
Ganoderma **6**:*773*
Gasteromycetidae **6**:*770*
Gastropoda **3**:*342*; **6**:*812*
Gavialidae **2**:*148*
Gecarcinidae **1**:*132*
Gelidium **8**:1127
Gempylidae **8**:*1060*
Genlisea **10**:1309
Gentianales **7**:*916*
Geomyidae **7**:*946*
Geophilomorpha **3**:*332*
Gerbillinae **7**:*946*
Gigantornis **1**:14
Gilpichthys **4**:449
Ginkgophyta **8**:*1138*

Giraffa
   camelopardalis **3**:*356*
   jumae **3**:361
Giraffidae **3**:*356*
Giraffoidea **3**:*356*; **7**:*964*;
   10:*1360*
Glires **4**:*476*
Gnathostomata **1**:*122*;
   3:*436*; **4**:*446, 460*;
   8:*1060, 1096*;
   9:*1238, 1280*
Gnetophyta **8**:*1138*
Gongora **6**:833
Gorilla **4**:*550*
   gorilla **1**:*88*
Gossypium **6**:752
Gramineae **6**:*744*–745
Gulo **10**:*1324*
Gymnocladium hossei
   8:1050
Gymnophiona **1**:*54*;
   6:*800*
Gymnospermae **1**:*36*;
   6:*824*; **7**:*916*;
   8:*1048, 1138*

H
Haliaeetus leucocephalus
   2:*220*, 221
Halictidae **4**:*526*
Heliamphora **10**:1301
Helianthus annus **6**:752
Heliozoa **2**:*164*
Hemiascomycetes **6**:*770*
Hemigaleidae **4**:*460*
Herpestidae **5**:*592, 664*;
   7:*928*
Herpesviridae **10**:*1314*
Hesperoidea **4**:*494*, 495
Heterobasidiomycetes
   6:*770*, 771
Heterodonta **3**:*342*, 343
Hexapoda **1**:*26*; **2**:*208*;
   3:*332*; **4**:*494*; **5**:*692*;
   10:*1344*

Hexaprotodon liberiensis
   4:*506*, 510
Hippocampinae **8**:*1096*
Hippopotamidae **4**:*506*
Hippopotamus amphibius
   4:*506*
Hippotraginae **10**:*1360*,
   1361
Hirudea **2**:*234*
Holothuroidea **9**:*1224*,
   1225
Homarus americanus
   5:*684*
Hominidae **1**:*88*, 89;
   4:*550*
Hominoidea **1**:*88*;
   4:*550*
Homo **4**:*550*, 551
   erectus **4**:*550*
   floresiensis **4**:551
   heidelbergensis **4**:*550*
   neanderthalensis **4**:*550*
   sapiens **1**:*88*; **4**:*550*
Homobasidiomycetes
   6:*770*, 771
Homoptera **5**:*692*
Hyaena hyaena **5**:*592*
Hyaenidae **5**:*592, 664*;
   7:*928*
Hyaeninae **5**:*592*
Hyatellidae **3**:*342*
Hydrobatidae **1**:*10*
Hydrochaeridae **7**:*898*
Hydrozoa **5**:*636*; **8**:*1084*
Hylobatidae **1**:*88*, 89;
Hylocereus undutus
   8:1050
Hymenomycetidae **6**:*770*
Hymenophyllaceae **2**:*276*
Hymenophyllum **2**:*276*
Hymenoptera **1**:*26*, 27;
   4:*494, 526*, 527
Hynobiidae **6**:801
Hypsignathus spp. **3**:*292*
Hyracoidea **2**:*244*

Hyracotherium **10**:*1421*,
   1424, 1426
Hystricidae **7**:*898*
Hystricognathi **7**:*898*,
   946

I
Ichneumonidae **1**:*26*
Ichthyostega **9**:1159, 1160
Indricothelium **8**:*1036*
Influenzavirus **10**:1314
Iniidae **2**:*188*
Insecta **1**:*26*; **2**:*208*;
   3:*332*; **4**:*494*, *526*,
   538; **5**:*692*; **8**:*1074*;
   10:*1344*
Ischnocera **5**:*692*, 693,
   694, 699
Ischyromys **7**:949
Isoptera **1**:*26*
Istiophoridae **8**:*1060*
Istiophorus
   albicans **8**:*1060*
   platypterus **8**:*1060*

J
Juncaceae **6**:*744*
Jynginae **10**:*1404*, 1405

K
Kinixys **9**:1268, 1273
Kogiidae **2**:*188*

L
Lacertilia **3**:*396*
Lagomorpha **4**:*476*
Lamiales **7**:*916*
Laminaria **8**:1131, 1133
Lasiocampidae **4**:*494*
Latimeria
   chalumnae **1**:*122*
   menadoensis **1**:*122*
Latimeriidae **1**:*122*
Lentibulariaceae
   10:*1300*–1301

Lentivirus **10**:1314
Lepdogalaxiidae **9**:*1281*
Lepidoptera **4**:*494*, 495
Lepidosirenidae **1**:*122*
Lepidosireniformes **1**:*122*
Leporidae **4**:*476*
Leptochariidae **4**:*460*
Leptochloa fusca **6**:748
Leptodactylidae **1**:*54*
Lepus **4**:*476*
Liliaceae **6**:*744*
Liliales **6**:*824*
Liliidae **6**:825
Lipotidae **2**:*188*
Lithobiomorpha **3**:*332*
Lontra canadensis **6**:*844*
Lophophora **8**:1052
Loxodonta **2**:*244*
   africana **2**:*244*
   cyclotis **2**:*244*
Lumbricidae **2**:*234*
Lumbricus terrestris **2**:*234*,
   235
Lutra lutra **6**:*844*
Lutrinae **10**:*1324*
Lybia tessellata **1**:135
Lycoperdon pyriforme **6**:*778*
Lycophyta **2**:*276*
Lycosa tarantula **9**:*1253*
Lyssavirus **10**:1314

M
Macaca **6**:*724*
Macroglossinae **3**:*292*
Macropodidae **5**:*648*, 649
Macropus **5**:*649*, 653
   giganteus **5**:*648*
   robustus **5**:*648*
   rufogriseus **5**:*648*
   rufus **5**:*648*
Maihuenia **8**:1050
Maihuenioideae **8**:*1048*,
   1049
Maihueniopsis glomerata
   8:1052

Majidae **1**:*132*

Malacostraca **1**:*132*, 133; **5**:*684*, 685

Malaxis paludosa **6**:832

Mallophaga **5**:692

Malus **1**:37
domestica **1**:*36*

Mammalia **1**:*88*; **2**:*188*, 244; **3**:*312*, *356*, *374*, *410*; **4**:*476*, *506*, *550*; **5**:*592*, *648*, *664*, *700*; **6**:*724*, *844*; **7**:*928*, *946*, *964*; **8**:*1032*; **10**:*1324*, *1360*, *1382*, *1418*

Mammillaria
hahniana **8**:1053
longimamma **8**:1052

Mandrillus **6**:*724*

Marmota **9**:*1208*

Marsilea **2**:280

Martes **10**:*1324*

Mastadenovirus **10**:1314

Mastigophora **1**:*20–21*; **2**:*164*

Mecoptera **4**:*538*

Megachilidae **4**:*526*

Megachiroptera **3**:*292*

Megadyptes antipodes **7**:*868*

Megalapteryx
didinus **6**:837
maximus **6**:837

Megalonychidae **3**:*312*; **9**:*1176*

Megaloptera **4**:*538*

Meganeuropsis **2**:217

Megaptera novaeangliae **3**:*374*

Megascolecidae **2**:234

Megatherium **3**:*320*; **9**:*1182*

Melinae **10**:*1324*

Meliponinae **4**:*526*

Meliursus ursinus **3**:*410*

Melocactus **8**:*1053*, 1056

Menopon gallinae **5**:693

Merostomata **9**:*1252*

Mesothelae **9**:*1252*, 1253

Mespilus **1**:*36*

Metatheria **1**:*88*; **5**:*648*; **6**:*724*, *844*

Metazoa **8**:*1084*–1085

Metridiidae **8**:1085

Microchiroptera **3**:*292*, 293

Microcladia **8**:1131

Mollusca **3**:*342*; **6**:*812*–813

Monera (Prokaryotae) **1**:*46*

Monocotyledonae **1**:*36*; **6**:*744*, *824*; **7**:*916*

Monodontidae **2**:*188*

Monognathidae **3**:*436*

Monoplacophora **3**:*342*; **6**:*812*

Monostroma **8**:1132

Monotremata **7**:*882*

Mucor **6**:771

Mucorales **6**:771

Mugiloidei **8**:*1060*

Muntiacus **7**:*964*

Muridae **7**:*946*, 947

Murinae **7**:947

Muscidae **4**:*538*, 539

Mustela **10**:*1324*, 1339

Mustelidae **3**:*410*; **6**:*844*, 845; **10**:*1324*, 1325, *1382*

Mustelinae **10**:*1324*, 1325, 1341

Myctophiformes **3**:439

Mygalomorphae **9**:*1252*

Myidae **3**:*342*

Myliobatidae **9**:1239

Myliobatiformes **9**:*1238*, 1239

Myoida **3**:*342*

Myomorpha **7**:*946*, 947

Myriapoda **8**:*1074*

Myrmecocystus **1**:*26*

Myrmecophaga **3**:*312*
tridactyla **3**:*312*

Myrmecophagidae **3**:*312*

Myrmicinae **1**:27

Mysticeti **2**:*188*; **3**:*374*

Myxine **4**:*446*, 448, 454
glutinosa **4**:*446*

Myxini **4**:*446*, 448, 450

Myxinidae **4**:*446*, 448

Myxiniformes **4**:*446*, 448

Myxinikela siroka **4**:*449*

Myxomycota **8**:*1126*

Myxophaga **10**:*1344*

N

Nautiloidea **6**:*812*, 813

Nautilus **6**:813

Navicula **2**:*166*, *167*, *168*

Nemalion multifidum **8**:1132

Nemamyxine **4**:*446*, 448

Nematocera **4**:*538*, 539

Neobalaenidae **3**:*374*

Neobatrachia **1**:*54*, 55

Neomyxine **4**:*446*, 448

Nepenthaceae **10**:*1301*

Nepenthales **10**:*1300*, 1301

Nepenthes **10**:*1304*–1305
kinabaluensis **10**:1303
rajah **10**:1301, 1308

Nephropidae **5**:*684*

Nephropoidea **5**:*684*, 685

Nesolagus **4**:*476*

Neuroptera **4**:*538*

Nicotiana
glauca **7**:*916*
tabacum **7**:925

Notacanthiformes **3**:*436*

Notomyotida **9**:*1224*

Notomyxine **4**:*446*, 448

Nyctimene spp. **3**:*292*

O

Obelia **5**:637

Ochotona **4**:*476*, 480

Ochotonidae **4**:*476*

Octopoda **6**:*812*

Octopodiformes **6**:*812*

Odobenidae **8**:*1110*

Odocoieus **7**:*964*

Odonata **2**:*208*; **5**:*692*

Odontoceti **2**:*188*; **3**:*374*

Odontomachus **1**:*26*

Okapi johnstoni **3**:*356*

Oligochaeta **2**:*234*

Oncorhynchus
clarkii **9**:1292
gilae **9**:1292
mykiss **9**:*1280*, 1284

Ophioglossaceae **2**:*276*

Ophioglossales **2**:*276*

Ophioglossum **2**:*276*

Ophiuroidea **9**:*1224*

Ophrys **6**:*832*

Opiliones **8**:*1074*, 1075

Opuntia **8**:1049, 1051, 1058

Opuntioideae **8**:*1048*, 1049, 1052

Orchidaceae **6**:*744*, *824*, 825

Orchidales **6**:*824*, 825

Orchidoideae **6**:*824*, 825, 826

Orchis **6**:825

Ornithorhynchidae **7**:*882*, 883

Ornithorhynchus anatinus **7**:*882*

Orthomyxoviridae **10**:1314

Orthopoxvirus **10**:1314

Oryctolagus cuniculus **4**:*476*

Oryza **6**:752

Osmeridae **9**:*1281*

Osteichthyes **1**:*122*, 123;
  3:*436*, 437; **4**:*460*;
  8:1061, *1096*, 1097;
  9:*1238*, *1280*, 1281
Osteolaemus **2**:*148*
Otariidae **8**:*1110*

P

Paenungulata **2**:*244*
Pakicetus **2**:192
Palaeudyptes antarcticus
  7:*872*
Paleognathae **6**:*834*
Paleoheterodonta **3**:*342*
Palinuroidea **5**:*684*, 685
Pan **4**:*550*
  paniscus **1**:*88*
  troglodytes **1**:*88*
Pandinus **8**:1075
Pandionidae **2**:*220*
Panthera **5**:665; **7**:932
  leo **5**:*664*, 665
  onca **5**:*664*
  pardus **5**:*664*
  tigris **5**:*664*
Papilionoidea **4**:*494*,
  495
Papillomavirus **10**:1314
Papio **6**:*724*
Papovaviridae **10**:1314
Parahyaena brunnea **5**:*592*
Paranthropus **4**:*550*
Pauropoda **3**:*332*
Paxillosida **9**:*1224*
Pecora **3**:*356*
Pediculidae **5**:693
Pediculus humanus **5**:*692*,
  693
Pelecanoididae **1**:*10*
Penicillium **6**:777
Pennales **2**:*164*
Pentalagus furnessi **4**:*476*
Perciformes **8**:*1060*,
  1061; **9**:*1280*
Pereskia diaz **8**:*1052*

Pereskioideae **8**:*1048*,
  1049
Perissodactyla **8**:*1032*,
  1033; **10**:*1360*,
  1361, 1366, *1418*
Petauristinae **9**:*1208*
Petrogale **5**:649
Petromyzontidae **4**:447
Petromyzontiformes
  4:447
Petunia **7**:*916*
Peyote **8**:1052
Phacochoerinae **4**:*506*
Phaeophyceae **8**:*1126*,
  1127
Phaeophyta **8**:*1126*, 1127
Phaethornithinae **5**:*580*
Phalaenopsis **6**:*825*
Phalangerida **5**:*648*, 649
Phalangeridae **5**:*648*
Phallus impudicus **6**:779
Phocidae **6**:*844*; **8**:*1110*
Phocoenidae **2**:*188*
Phoebetria **1**:11
  fusca **1**:*10*
Phoridae **4**:*538*
Phthiraptera **5**:*692*
Phthiridae **5**:693
Phthirus pubis **5**:693,
  695
Physeteridae **2**:*188*
Phytophthora infestans
  6:770
Picidae **10**:*1404*, 1405
Piciformes **10**:*1404*
Picinae **10**:*1404*
Picoides **10**:1405
Picornaviridae **10**:1314
Picumninae **10**:*1404*,
  1405
Pilobolus **6**:778–779
Pilosa **3**:*312*; **9**:*1176*
Pinaceae **8**:*1138*, 1139
Pinnipedia **3**:*410*;
  8:*1110*; **10**:*1382*

Plantae **6**:*744*, *824*;
  7:*916*; **8**:*1048*,
  1138
Platanistidae **2**:*188*
Platybelodon **2**:247
Platycerium **2**:*276*
Platyrrhini **1**:*88*; **4**:*550*;
  6:*724*
Platysternidae **9**:*1194*,
  1266
Plectomycetes **6**:770, 771
Plethodontidae **6**:*800*,
  801
Pleurodira **9**:*1194*, *1266*
Pleurosigma **2**:166
Pogonomyrmex **1**:*26*, 27
Polemoniales **7**:*916*
Polychaeta **2**:*234*
Polyphaga **10**:*1344*, 1345
Polyplacophora **3**:*342*;
  6:*812*
Polypodiaceae **2**:*276*
Polysiphonia **8**:*1134*
Pomoideae **1**:*36*
Ponerinae **1**:27
Pongo **4**:*550*
  abelii **1**:*88*
  pygmaeus **1**:*88*
Pontoporidae **2**:*188*
Porphyra **8**:*1126*, 1127,
  1134, 1135–1136,
  1137
Portunidae **1**:*132*
Postelsia palmaeformis
  8:1131
Potoroidae **5**:649
Poxviridae **10**:1314
Primates **1**:*88*; **4**:*550*;
  6:*724*
Priscacara liops **9**:*1158*
Pristiformes **9**:*1238*
Pristina leidyi **2**:243
Proboscidea **2**:*244*
Procellariidae **1**:*10*
Procellariiformes **1**:*10*

Procoptodon goliah **5**:652
Procyonidae **3**:*410*
Prokaryotae (Monera)
  1:*46*
Proscyllidae **4**:*460*
Proteidae **6**:801
Proteles cristatus **5**:*592*
Protelinae **5**:*592*
Proteobacteria **1**:*46*, 47
Protista **2**:*164*; **8**:*1126*
Protopteridae **1**:*122*
Protosciurus **9**:1215
Prunus **1**:*36*, 38, 44
Pseudoscorpiones
  8:*1074*, 1075
Pseudotriakidae **4**:*460*
Psilophyta **2**:*276*
Pteridium **2**:*276*, 279
Pteridophyta **2**:*276*
Pteriomorpha **3**:*342*
Pterocactus **8**:*1049*
Pteromyinae **9**:*1209*
Pteronura brasiliensis **6**:*844*
Pteropodinae **3**:*292*
Pteropus spp. **3**:*292*, 293
Puccinia **6**:771
Pycnogonida **9**:*1252*
Pycnopodia helianthoides
  9:*1224*
Pygoscelis **7**:*868*, 873, 876
Pyracantha **1**:*36*
Pyrenomycetes **6**:*770*,
  771
Pyrrophyta **8**:*1126*
Pyrus **1**:37
  communis **1**:*36*
Pythonidae **3**:*396*
Pyxis **9**:1268

R

Radiolaria **2**:*164*
Rajiformes **9**:*1238*
Ramphomicron **5**:*582–583*
Rana **1**:55
  berlandieri **1**:*54*

Rana (continued)
  catesbeiana **1**:*54*
  esculenta **1**:*54*
  temporaria **1**:*54*
Rangifer tarandus **7**:*964*
Ranidae **1**:*54*, 55, *56*
Raphidoptera **4**:*538*
Rattus **7**:947
Rebutia **8**:1059
Reptilia **2**:*148*; **3**:*396*;
  5:*624*; **9**:*1266*
Retropinnidae **9**:*1281*
Retroviridae **10**:1314
Rhabdoviridae **10**:1314
Rhamnales **1**:*36*, 37
Rheiformes **6**:*834*
Rhinastus latesternus
  10:*1345*
Rhinobatiformes **9**:*1238*
Rhinoceros
  sondaicus **8**:*1032*
  unicornis **8**:*1032*
Rhinocerotidae **8**:*1032*;
  10:*1418*
Rhinodermatidae **1**:*54*
Rhizanthella gardneri
  6:*827*, *828*
Rhizopoda **1**:21, 22;
  8:*1126*
Rhizostomae **5**:*636*,
  638
Rhodophyceae **8**:*1126*
Rhodophyta **8**:*1126*,
  1127
Rhyacotritonidae **6**:801
Rhyncophthirina **5**:*692*,
  693
Rodentia **4**:*476*; **6**:727;
  7:*898*, *946*; **9**:*1208*,
  1209
Rodhocetus **2**:192
Romerolagus diazi **4**:*476*
Rosa **1**:37
Rosaceae **1**:*36*, 37
Rosales **1**:*36*, 37

Rosidae **1**:*36*, 37;
  10:*1300*
Rousettus spp. **3**:*292*
Rubivirus **10**:1314
Ruminantia **3**:*356*;
  4:*506*; **7**:*964*

S

Saccharum officinarum
  6:*744*
Saccopharyngidae **3**:*436*
Saccopharyngiformes
  3:*436*, 437
Saccopharyngoidei **3**:*436*
Saccopharynx **3**:438–439,
  441, 442, 444
Sagartia troglodytes **8**:1094
Sagittariidae **2**:*220*
Salamandridae **6**:*800*,
  801, 802
Salangidae **9**:1281
Salmonidae **9**:*1280*, 1281
Salmoniformes **9**:*1280*,
  1281
Salmo salar **9**:*1280*
Salticidae **9**:1260
Salvelinus fontinalis **9**:*1280*
Samotherium boissieri **3**:361
Sarcodina **1**:*20*, 21; **2**:*164*
Sarcophagidae **4**:*538*
Sarcopterygii **1**:*122*, 123;
  3:*436*; **8**:*1060*;
  9:*1280*
Sarracenia **10**:*1300*
Sarraceniaceae **10**:*1300*,
  1301
Saturnidae **4**:*494*
Sauria **5**:*624*
Saurosuchus imperator
  2:154
Scandentia **6**:727
Scaphopoda **3**:*342*; **6**:*812*
Schizophora **4**:*538*, 539
Sciuridae **9**:*1208*, 1209
Sciurinae **9**:*1208*, 1209

Sciurognathi **4**:547;
  7:*898*, *946*; **9**:*1208*,
  1209
Sciurus **9**:*1208*, 1209
Scleractinia **8**:*1084*
Sclerocactus papyracanthus
  8:1053
Scolioidea **4**:*526*
Scolopendramorpha
  3:*332*
Scombroidei **8**:*1060*,
  1061
Scorpiones **8**:*1074*;
  9:*1252*
Scrophulariales **7**:*916*
Scutigeromorpha **3**:*332*
Scyliorhinidae **4**:*460*
Scyphozoa **5**:*636*; **8**:*1084*
Selachii **9**:*1238*
Semaeostomae **5**:*636*,
  638
Sepioida **6**:*812*, 813
Sepiolida **6**:813
Sequoidadendron giganteum
  8:*1138*, 1139
Serpentes **3**:*396*; **5**:*624*
Sigmodontinae **7**:*946*,
  947
Simplexvirus **10**:1314
Siphonaptera **4**:*538*
Sirenia **2**:*244*; **5**:*700*
Sivatherium **3**:361
Skrjabingylus nasicola
  10:1332
Solanaceae **7**:*916*, 925
Solandra **7**:925
Solanum **7**:917
  lycopersicum **7**:*916*
  tuberosum **7**:*916*
Solenidae **3**:*342*
Solpugida **8**:*1074*, 1075
Spartina **6**:*744*, 745, 750
  alterniflora **6**:*744*, 750
  anglica **6**:*744*, 750, *751*
  maritima **6**:*744*, 750

Spermophilus **9**:*1208*,
  1212
Spheniscidae **7**:*868*
Sphenisciformes **7**:*868*
Spheniscus **7**:*868*
Sphenophyta **2**:*276*
Sphingidae **4**:*494*, 495
Sphingoidea **4**:495
Sphyraenoidei **8**:*1060*
Sphyrnidae **4**:*460*
Spinulosida **9**:*1224*
Spiranthoideae **6**:*824*
Spirochaetes **1**:47
Spirulida **6**:813
Squamata **3**:*396*; **5**:*624*;
  9:*1194*, *1266*
Staphylinidae
  10:1347–1348
Stauromedusae **5**:*636*
Stenocereus
  eruca **8**:1050
  thurberi **8**:1050
Stramenopiles **2**:*164*
Strepsiptera **4**:*538*
Strepsirrhini **4**:*550*
Strigamia maritima **3**:335
Struthio camelus **6**:*834*
Struthioformes **6**:*834*
Stylasterias forreri **9**:1227
Suidae **4**:*506*
Suiformes **4**:*506*; **7**:*964*;
  10:*1360*
Suinae **4**:*506*
Syconycteris spp. **3**:*292*
Sylvilagus **4**:*476*
Symphata **4**:*526*
Symphyla **3**:*332*
Symphyta **1**:*26*
Syngnathidae **8**:*1096*
Syngnathiformes **8**:*1096*,
  1098
Syngnathinae **8**:*1096*,
  1098
Syngnathus **8**:1109
Syrphidae **4**:*538*

T

Tachinidae **4**:*538*
Tachyglossidae **7**:*882*, 883
Tachyglossus aculeatus **7**:*882*
Tamandua **3**:*312*
Tamias **9**:*1208*
Tapiridae **8**:*1032*; **10**:*1418*
Tarsii **4**:*550*
Taurotragus **10**:*1364*
Taxaceae **8**:1139
Taxodiaceae **8**:*1138*, 1139
Teredinidae **3**:*342*
Termitomyces titanicus **6**:775
Testudinata **9**:*1194*
Testudinidae **9**:*1194*, *1266*
Testudo **9**:1273
Teuthoida **6**:*812*, 813
Thaumastochelidae **5**:*684*
Theales **10**:*1300*, 1301
Thelymitra spiralis **6**:828
Theraphosidae **9**:*1252*, 1253
Thermodesulfobacteria **1**:47
Thymollus arcticus **9**:*1280*
Tiburonia granrojo **5**:638
Tinamiformes **6**:*834*
Togaviridae **10**:1314
Torpediniformes **9**:*1238*
Tragulina **3**:*356*
Tremarctos ornatus **3**:*410*
Tremella **6**:771
Trichechidae **5**:*700*
Trichechus
  inunguis **5**:*700*
  manatus **5**:*700*
  senegalensis **5**:*700*
Trichoptera **4**:*538*
Tridacna **3**:343
  gigas **3**:*345*, 347
  maxima **3**:*343*, 344
Tridacnidae **3**:*342*

Trikidae **4**:*460*
Triphyophyllum peltatum **10**:1301
Triticum **6**:751, 752
Triturus **6**:801
Trochilidae **5**:*580*, 581
Trochiliformes **5**:*580*, 581
Trochilinae **5**:*580*, 581
Tuber **6**:775
Tulasnellales **6**:771
Tulerpeton **9**:1160
Tylopoda **4**:*506*; **7**:*964*

U

Ulmaridae **5**:*636*, 638
Ulva lactuca **8**:*1126*
Ulvophyceae **8**:*1126*
Uniramia **2**:*208*; **3**:*332*; **4**:*494*; **9**:*1252*
Uredinales **6**:771
Urodela **6**:*800*
Urolophus **9**:1247
Ursavus **3**:415
Ursidae **3**:*410*; **6**:*844*
Ursus
  americanus **3**:*410*
  arctos **3**:*410*
  malayanus **3**:*410*
  maritimus **3**:*410*
  thibetanus **3**:*410*
Ustilaginales **6**:771
Ustilago **6**:771

V

Valvatida **9**:*1224*
Vampyromorpha **6**:813
Vampyromorphoida **6**:*812*
Velatida **9**:*1224*
Veneroida **3**:*342*
Vermilingua **3**:*312*
Vertebrata **1**:*122*; **2**:*148*, 188, 220, 244; **5**:*700*; **7**:*868*, 898

Vesicomyidae **3**:*342*
Viveridae **5**:*592*, *664*; **7**:*928*
Vombatiformes **5**:*648*

W

Woodsiaceae **2**:*276*

X

Xenarthra **3**:*312*; **9**:*1176*, 1177, 1181, 1191
Xiphiidae **8**:*1060*

Z

Zaglossus **7**:*882*
Zea mays **6**:*749*, 753
Zingiberidae **6**:825
Ziphiidae **2**:*188*
Zygomycota **6**:*770*, 771
Zygoptera **2**:*208*

Comprehensive index

A

aardvark
  brain **3**:328
  sense of smell **3**:328
  teeth **3**:321
aardwolf **5**:*592*, 593, 598,
    599, 603, 604
  fur markings
    5:*594–595*
  skull **5**:*597*
  teeth **5**:*597*
  toes **5**:*594*
  tongue **5**:*599*
abductor muscles, of
    coelacanths **1**:*127*
abomasum **3**:*370*; **7**:*979*;
    10:1376, 1377,
    1380–1381
Acanthaster planci **9**:*1224*
Acanthostega **9**:1160
Acarina **8**:*1074*–1075;
    9:*1252*
Accipitridae **2**:*220*
acetabula **3**:300
acetylcholine **1**:121;
    4:565; **6**:760–761
acontia **8**:1093
Acrosiphonia coalita
    8:1131
ACTH
    (adenocorticotropic
    hormone) **2**:271
Actinaria **8**:*1084*, 1085
actin filaments **6**:754,
    756, 757–*759*, 760,
    767, 768
Actinopoda **1**:21, 22
Actinopterygii **3**:*436*,
    437; **8**:*1060*, 1061,
    1096, 1097; **9**:*1174*,
    1280, 1281
action potentials
    6:783–785, 786;
    10:1310
Addison's disease
    5:623

adductor muscles
  of clams **3**:349, *350*
  of coelacanths **1**:127
adenosine diphosphate
    see ADP
adenosine triphosphate
    see ATP
Adenoviridae **10**:1314
Adephaga **10**:*1344*,
    1345, 1357
Adetomyrma **1**:*26*, 33
ADP (adenosine
    diphosphate) **6**:759
adrenal glands **2**:268,
    271–272; **4**:570
aedeagus **5**:699
  of dragonflies **2**:213,
    218, 219
  of weevils **10**:1357
Aepyornis maximus **6**:837
Aeshnidae **2**:*208*
Africa
  aardwolves **5**:593
  antelope **10**:*1376*
  chameleons **5**:624
  hyenas **5**:593
  lions **5**:664
  manatees **5**:700
  ostriches **6**:835
  pangolins **9**:*1181*
  porcupines **7**:900, 901,
    902, 907
  scaly-tailed flying
    squirrels **9**:1213
  scorpions **8**:1075
agar **8**:1127
Agaricaceae **6**:*770*, 771
Agaricales **6**:*770*, 771
Agaricus
  bisporus **6**:*770*, 771
  campestris **6**:*770*
Agnatha **3**:*436*; **4**:446,
    460; **8**:*1060*, *1096*;
    9:*1238*, *1280*
AIDS **5**:616
  see also HIV

Ailuropoda melanoleuca
    3:*410*
air sacs (parabronchi)
  of birds **3**:307;
    8:1020, 1028
  albatrosses **1**:*17*
  eagles **2**:231; **8**:*1019*
  hummingbirds **5**:588
  ostriches **6**:841
  woodpeckers **10**:1414
Aizoaceae **8**:1049
Alaria esculenta **8**:*1130*
Alaska, wolves **10**:1385
alates **1**:34
albatross **1**:**10–19**
  bill **1**:*11, 13*, 18
  circulatory and
    respiratory systems
    1:11, *17*
  digestive and excretory
    systems **1**:11, *18*
  external anatomy **1**:*11,
    12–13*
  flight **1**:*12*–13, *15*
  gray-headed **1**:*10*
  Hawaiian black-footed
    1:16
  Laysan **1**:*10*, 16, *19*
  long journeys **1**:13
  muscular system **1**:11,
    15
  reproductive system
    1:11, *19*
  royal **1**:*10*
  senses **1**:16
  short-tailed **1**:*10*
  skeletal system **1**:11,
    14
  sooty **1**:*10*, 11
  taxonomy **1**:10
  vision and eyesight
    1:*13*, 16
  wandering **1**:*10, 12,
    13, 14, 15*
  yellow-nosed **1**:*10*
Albuliformes **3**:*436*

alcohol dehydrogenase
    4:568
alderfly **4**:*538*
Aldrovanda **10**:1301
algae
  brown **8**:*1126*, 1127,
    1129, *1130*, 1131,
    1133
  green **2**:164; **8**:*1126*,
    1127, 1131
  life cycles **8**:1134–1135
  red **8**:*1126*, 1127,
    1131, 1133
  reproduction **8**:*1134*,
    1135–1136
  see also seaweed
algae, blue-green *see*
    cyanobacteria
alimentary canal,
    mammalian **6**:*741*
Alismatidae **6**:825
alleles **1**:76–79
allergies **5**:*613*, 615–616
Alligatoridae **2**:*148*
alternation of
    generations
  ferns **2**:*283*–284;
    7:*993*; **8**:1137
  hydrozoans **7**:986
  jellyfish **5**:646;
    7:986–987
  seaweeds **8**:*1134*–1135
  water fleas **7**:*989*
altitudes, high, living at
    4:566
altricial young **4**:493
altruism, in wolves
    10:1401
Alveolata **2**:*164*
alveolates **2**:*164*
alveoli **4**:*567*; **5**:677;
    7:911; **8**:*1012*, 1020
Amanita
  muscaria **6**:771
  phalloides **6**:771
  virosa **6**:771

Amanitaceae **6**:*770*
ambergris **2**:205
Amblycera **5**:*692–693*
ambulacral groove **9**:*1227*
ambulacral muscles **9**:*1231*
Ambystoma **6**:801, 811
Ambystomatidae **6**:*800*, 801
amebocytes **1**:138
amino acids **7**:984
ammonite **6**:*815*
Ammonoidea **6**:815
amoeba (ameba) **1**:*20–25*; **2**:*164*, 166; **8**:*1126*
  external anatomy **1**:*21*, 22
  internal anatomy **1**:21, 23
  reproductive system **1**:21, *24–25*
Amoeba proteus **1**:*20*, 22
Amoebidae **1**:*20*
Amphibia **1**:*54*; **6**:*800*
Amphisloaemia **5**:*624*
Amphiumidae **6**:801
ampullae of Lorenzini **4**:*470*, *471*; **6**:*787*; **9**:*1247*
amygdala **4**:565; **5**:709; **6**:798
amylase **2**:173; **4**:*569*
Anabaena **1**:*49*
anadromous species, lampreys **4**:448
anal glands
  of pumas **7**:933
  of weasels **10**:1335, 1336
  of wolves **10**:1399
anal sacs, of turtles **9**:1204, 1205
anaphylatoxins **5**:611
Anax imperator **2**:*208*

Androctonus **8**:1081
anemia, sickle-cell **1**:*78*, 79
anemone
  coral-like **8**:*1084*
  tube **8**:*1084*
  see also sea anemone
angelfish, queen, changing sex **7**:998
Angiospermae **1**:*36*; **6**:*824*; **7**:*916*, 917; **8**:*1048*, *1138*
angiosperms **1**:*36*, 37; **6**:*744*, *824*; **7**:*916*, 917; **8**:*1048*, *1138*; **10**:*1300*
  reproductive structures **7**:1000–*1003*
angler fish, deep-sea **3**:440, *444*
Anguilliformes **3**:*436*
Anisoptera **2**:*208*
Anisozygoptera **2**:*208*
anisozygopterans **2**:*208*, 209
ankylosing spondylitis **5**:622
annelid worms **2**:234
  hormones **2**:269
  mouthparts **2**:236
  skeletal system **9**:1163–1164
Anomalodesmata **3**:*342*
Anomaluridae **9**:*1208*
Anomura **1**:*132*, 133, 135
anomurans **1**:*132*, 133
Anoplura **5**:*692*, 693, 694
ant **1**:*26–35*; **4**:*526*, *538*
  anatomy
    external **1**:27, *28–29*
    internal **1**:27, *30*
  black garden **1**:*29*
  carpenter **1**:*28*, *31*, *34*

castes **1**:29
chemical
  communication **1**:31
  differs from termites **1**:29
  digestive and excretory systems **1**:27, *32–33*
  dracula **1**:*26*, 27, 33
  as gardeners or farmers **1**:*33*
  harvester **1**:*26*, 27, 32
  honeypot **1**:*26*, 27, 33
  king and queen **1**:35
  leaf-cutter **1**:*26*, 27, *33*, 35; **6**:773
  nervous system **1**:27, *31*
  ovipositor function **1**:35
  reproductive system **1**:27, *34–35*
  respiration **1**:30
  soldier **1**:29
  trap-jaw **1**:*26*, 27, 28
anteater **3**:*312*, 313; **9**:*1176*, 1177
  digestive system **9**:1191
  giant *see* giant anteater
  silky **3**:*312*, 313, *315*
    claws **3**:316
    skull **3**:*319*
    tail **3**:317, *324*
    teeth **3**:321
  spiny *see* echidna
antelope **3**:*356*, 357; **7**:*964*, 967–968; **10**:*1360*
  dwarf **10**:*1360*
  four-horned **10**:*1365*, *1367*, *1376*
  grazing **10**:*1360*
  limbs **10**:*1369*, 1371
  spiral-horned **10**:*1360*, *1364*
  toes **10**:*1366*, 1367

antennules **1**:135
antheridia **2**:283, 284
anthers **1**:43; **6**:751; **7**:*927*
Anthophoridae **4**:*526*
Anthozoa **5**:*636*; **8**:*1084*
anthozoans **5**:*636*, 638, 639; **8**:*1084*
Anthribidae **10**:*1344*
antibodies **5**:608, 609, *612–614*, 620
  autoantibodies **5**:622
  see also immunoglobulins
antibody-dependent cell-mediated cytotoxicity (ADCC) **5**:613, 615
antidiuretic hormone (ADH) **2**:268; **4**:571
antigenic drift **5**:621
antigenic shift **5**:621
antigens **5**:612, *613*
Antilocapridae **10**:1371
Antilopinae **10**:*1360*
antioxidants **1**:72
Antipatharia **8**:*1084*
antlers **7**:*965*, *966*, *969*, *971–972*, 974, 975, *982*; **9**:*1171*
  females with **7**:982
Anura **1**:*54*, 60; **6**:*800*
aorta **1**:115
Apatelododae **4**:*494*
ape **1**:*88*, 89; **4**:*550*; **6**:*724*
  great **1**:*88*, 89
Aphelocheirus **8**:*1018*
aphid **1**:*33*; **5**:*692*
Aphonopelma **9**:*1265*
Aphyllophorales **6**:771
Apiales **1**:*36*, 37
apical complex **1**:21
Apicomplexa **1**:*20*, 21
Apidae **1**:*26*; **4**:*526*, 527

Apinae **4**:*526*
Apis **4**:527
  andreniformis **4**:*526*
  cerana **4**:*526*
  dorsata **4**:*526*
  florea **4**:*526*
  koschevnikavi **4**:*526*
  laboriosa **4**:*526*
  mellifera **4**:*526*
Aplacophora **3**:*342*;
  6:*812*
Aplysia **6**:819
Apocrita **1**:*26*, 27; **4**:*526*,
  527
apocritans **4**:527
apodemes **2**:214; **9**:1258
Apodiformes **5**:*580*
Apodimorphae **5**:*580*
Apoidea **4**:*526*, 527
apoikogenic
  development
  8:*1082*–1083
apomixis **7**:*988*, 989
apoptosis **1**:87; **5**:610
Apostasioideae **6**:*824*,
  825
appendix **1**:102; **2**:182;
  4:569
apple tree **1**:**36–45**
  anatomy
    external **1**:37,
     38–39
    internal **1**:37,
     40–42; **9**:*1161*
  reproductive system
    1:37, *43–45*;
    7:*1001*
apricot **1**:44
Aptenodytes **7**:*868*
Aquificae **1**:*46*, 47
Arachnida **8**:*1074*;
  9:*1252*, 1253
arachnids **8**:*1074*;
  9:*1252*, 1253
  skeletal system
    9:1166–1167

Araneae **8**:*1074*; **9**:*1252*,
  1253
Araneomorphae **9**:*1252*,
  1253
araneomorphs **9**:*1252*
Araucaria **8**:1139
Araucariaceae **8**:1139
Archaea **1**:*46*
Archaeobatrachia **1**:*54*,
  55
Archaeopteryx **9**:*1160*
  brain **6**:796
archegonia **2**:283, 284,
  285
Archilochus colubris **5**:*580*
Archostemata **10**:*1344*,
  1345
Arctic Ocean, gray
  whales **3**:375
Arctodus **3**:415
Arecidae **6**:825
Arenaviridae **10**:1314
Arenavirus **10**:1314
areoles **8**:1052
Arethuseae **6**:825
Ariocarpus **8**:1052, 1055
Arizona, cacti **8**:*1051*
armadillo **3**:*312*, 313;
  9:*1176*, 1177, 1181
  brain **3**:328
  claws **3**:316
  exoskeleton **9**:1169
  giant **3**:321; **9**:1185
  nine-banded **9**:*1169*
  sense of smell **3**:328
  teeth **3**:321
  yellow **9**:*1181*
Armillarea bulbosa **6**:772
arteries **1**:*118*; **4**:487
  hardening of the
    4:567
arterioles **1**:118, 119, 121
arteriosclerosis **4**:567
arthritis, rheumatoid
  5:622, *623*
Arthrobotrys **6**:774

Arthropoda **1**:*26*, *132*;
  2:*208*; **3**:*332*;
  4:*494*, *526*, *538*;
  5:*684*, *692*; **8**:*1074*;
  9:*1252*
arthropods **1**:*26*, *132*;
  2:208; **3**:*332*;
  4:*494*, *526*, *538*;
  5:*684*, *692*; **8**:*1074*;
  9:*1252*
  digestive system **2**:179
  exoskeleton **9**:1165,
    1166–1167
  molting **2**:267
  mouthparts **2**:175
  nervous systems **6**:793
Artiodactyla **3**:*356*;
  4:*506*; **7**:*964*;
  8:*1032*; **10**:*1360*,
  1361, *1418*
artiodactyls *see* ungulates,
  even-toed
Arvicolinae **7**:*946*
Aschiza **4**:*538*
asci **6**:771, 774
Ascomycetes **6**:776, 779
Ascomycota **6**:770, 771,
  775
Asia
  ground squirrels
    9:*1212*
  pangolins **9**:*1181*
  pikas **4**:480
  porcupines **7**:900, 901
  tigers **5**:668
  wolverine **10**:*1330*,
    1331
Asplenium bulbiferum
  2:285
ass **8**:*1032*
  African **10**:*1418*
  Asiatic **10**:*1418*
  horse-ass hybrids
    10:1437
Asterias forbesi **9**:*1237*
Asteridae **7**:*916*

Asteroidea **9**:*1224*
atherosclerosis **1**:*118*
Athyrium **2**:*276*
Atlantic Ocean, stingrays
  9:*1239*
atlas bone (vertebra)
  4:557; **6**:731;
  9:1173; **10**:1391
ATP (adenosine
  triphosphate) **1**:72,
  73; **6**:741
  powering muscles
    6:*759*, 760, 761
atrioventricular (AV)
  node **1**:116–117;
  6:767
Atropa belladonna **7**:*916*
atropine **7**:*925*
Atta **1**:*26*
attacins **5**:608
Attelabidae **10**:*1344*
auk, wing bones **7**:*874*
Aurelia aurita **5**:*636*
Auricularia **6**:771
Australia
  crown-of-thorns
    starfish **9**:1235
  dingos **7**:885
  echidnas **7**:883
  giant sea anemones
    8:1086
  kangaroos **5**:648
  monotremes and
    marsupials **7**:885
  pitcher plants
    10:*1304*
  sea anemones
    8:*1091*
Australopithecus **4**:551
autoimmunity **5**:608,
  622, *623*
autoregulation **1**:121
autosomes **7**:990
autotomy **2**:237; **5**:629
autotrophs **1**:66
  bacteria as **1**:52

auxin **2**:266
auxospores **2**:171
Aves **1**:*10*; **2**:*220*; **5**:*580*;
    6:*834*; **7**:*868*, 869;
    10:*1404*
avian flu virus *see* bird flu
    virus
awns **6**:751, 753; **10**:1386
axis bone (a vertebra)
    4:*557*; **9**:1173
axons **6**:782, *785*
    giant, of squid **6**:783
axopods **2**:167
Azolla **2**:280

B
babirusa **4**:*506*
baboon **6**:*724*, 725
    hands and feet **6**:*728*
Babyrousinae **4**:*506*
Bacillariophyta **2**:*164*
Bacillus **1**:53
backswimmer
    8:*1017–1018*, 1025
bacteria **1**:**46–53**
    aerobic and anaerobic
        1:52
    anatomy
        external **1**:47, *48–49*
        internal **1**:47, *50–51*
    autotrophic **1**:52
    cellulose-digesting
        2:182
    digestive and excretory
        systems **1**:47, *52*
    enteric **1**:47
    Gram-positive and
        Gram-negative
        1:*48–49*
    heterotrophic **1**:52
    immune system and
        5:606–614
    lipopolysaccharide
        (LPS) coat **5**:610
    nitrogen-fixing **1**:47;
        2:172

pathogenic **5**:606, 610
reproductive system
    1:47, *53*
    spores **1**:*53*
bacteriophage **10**:*1320,
    1321, 1322*
baculum **7**:929; **8**:*1124*;
    10:*1341*, 1342,
    1383, 1400
badger **10**:*1324*, 1333
    American **10**:1331,
        1333
    European/Eurasian
        10:1331, 1333,
        1339
Balaenoptera
    musculus **3**:*374*
    omurai **3**:*374*
    physalus **3**:*374*
Balaenopteridae **3**:*374*
balance, sense of **6**:787,
    788–789
baleen **2**:175, 189; **3**:375,
    377, 378, *392, 393*
Bangiophycidae **8**:1127
barbet **10**:*1404*
bark **1**:38
barnacle **3**:*378*
barracuda **8**:*1060*
basidia **6**:771
Basidiomycetes **6**:776,
    779
Basidiomycota **6**:*770*
Basilosaurus **2**:192
basophils **1**:*108*, 109;
    5:611
bat **3**:292
    blossom **3**:*292*
    circulatory system
        1:*106, 115*
    fruit *see* fruit bat
    lesser long-nosed
        8:*1058*
    long-tongued **7**:*1002*
    pollination by **7**:*1002*;
        8:*1058*

vampire, teeth **2**:*174*
    wings **9**:1175
bataleur **2**:224
bathypelagic zone, gulper
    eels **3**:439
Batoidea **9**:*1238*, 1239
B-cells *see* lymphocytes
beak
    bird **9**:1170
    octopus **2**:*175*
bean **1**:*36*
bear **3**:*410*, 411; **5**:*592,
    664*; **6**:*844*; **7**:*928*;
    10:*1324*
    American black **3**:*410*,
        411, *415*
    Asian black **3**:*410*, 411,
        414
    brown **3**:*410*, 411, *415*
    dawn **3**:415
    grizzly *see* grizzly bear
    Kodiak **3**:411
    polar **3**:*410*, 411, *415*
        hibernation **3**:*425*
        skeleton **3**:*418*
    short-faced **3**:*415*
        skull **3**:*417*
    sloth **3**:*410*, *414*
        skull **3**:*419*
    spectacled **3**:*410*, *418*
        skeleton **3**:*418*
    sun **3**:*410*, *414*
beaver, tail **7**:955
bee **1**:*26*, 27; **4**:*494*, 526,
    538
    Africanized (killer)
        4:*530–531*
    carpenter **4**:*526*, 531
    euglossine/orchid
        6:833
    leafcutter **4**:*526*
    long-tongued **4**:*526*,
        527
    plasterer **4**:*526*
    stingless **4**:*526*, 527
    sweat **4**:*526*

wings **2**:211
see also honeybee
beetle **4**:*494*, *538*; **5**:*692*;
    10:*1344*
    "ancient" **10**:*1344*
    antennae **10**:*1347*,
        1348
    bombadier, toxins
        2:275
    click **10**:*1349*, *1350*
    cucujo **10**:*1354*
    evolution **10**:*1348*
    external anatomy
        10:*1346–1348*
    goliath **8**:*1021*
    ground **10**:*1344*
    longhorn **8**:1148
    metamorphosis
        10:1358
    rove **10**:*1347–1348*
    skiff **10**:*1344*
    wings **2**:211
Belidae **10**:*1344*
belladonna **7**:*916, 925*
beluga **2**:*188*, 190
    communication **2**:193,
        201
    facial expressions **2**:193
bends, the **2**:203; **8**:1121
Bergmann's rule **7**:870,
    958
Bernoulli effect **8**:1026
bettong **5**:649, 656
biceps femoris
    elephant **2**:255
    human **4**:561
    rhino **8**:1041
    seal **8**:1111
biceps muscles **4**:*560*;
    6:734, *754*; **8**:1040
biddy **2**:*208*
bile **2**:182, 204; **5**:590;
    6:741; **10**:1399
bilharziasis **7**:987
billfish **8**:1061, 1062,
    1064

billfish *(continued)*
  migration **8**:1073
  swimming
    8:1066–1067
binary fission **1**:24, 53,
    85; **7**:986
  longitudinal **8**:1094
biological control, of
    rabbits **4**:488
bioluminescence **3**:439,
    440, 444;
    10:*1348–1351*,
    1354
biotrophs, fungi as
    6:774
bipedalism **1**:95;
    4:*552–553*, 556,
    558, 561
  chimpanzees **1**:*94*, 97
bird flu virus **5**:*614*;
    10:*1319*
birds **1**:*10*; **2**:220; **5**:*580*;
    6:*834*; **7**:*868*, 869;
    10:*1404*
  beaks **9**:1170
  bones **1**:14; **9**:1168
  brain **2**:229; **6**:*795*
  calcium regulation
    2:268
  digestion by **2**:178
  embryo **7**:*1005*
  evolution from
    dinosaurs **9**:*1160*
  excretion by **2**:186
  flight **1**:13
  hearing by **9**:1170
  of prey **2**:220
  respiratory systems
    8:*1019*
    ventilation **8**:1028
  skeletal system **9**:1160,
    1175
  jaw **9**:1170
  song **7**:*999*
bird's nest fungi **6**:771,
    778

bivalves **3**:*342*
  catch muscles **6**:768
  circulatory system
    1:107
  heterodont **3**:*342*, 343
  nervous system **3**:*353*
Bivalvia **3**:*342*; **6**:*812*
blackberry (bramble)
    1:38, 39
blackfish **2**:*188*
bladderwort **10**:1305
blastocoel **7**:1004
blastocysts **8**:1124
blastula **7**:1004
blood **1**:*108–109*
  clotting **1**:*119*
  sugar levels **2**:*272*
blood-brain barrier
    6:795
blood cells **1**:*108–109*
  platelets **1**:108, *109*,
    119
  red (erythrocytes)
    1:*108–109*, 119;
    4:*521*; **8**:*1029*
  white (leukocytes)
    1:*108*, *109*; **5**:611
  see also basophils;
    eosinophils;
    lymphocytes;
    monocytes;
    neutrophils
blood fluke, asexual
    reproduction **7**:987
blood pressure **1**:*121*
bloodworm **4**:547
Blossfeldia lilputana
    8:*1050*
blowfly **4**:*538*, 539
  sheep, maggots **4**:548
blowholes **2**:195, *196*,
    202, *203*; **3**:380,
    386, 391; **8**:*1014*
blubber **2**:191, 192, 198
  seal **8**:1113–1114
  whale **3**:377, 378, *391*

bobcat **7**:945
Boidae **3**:*396*
Boletaceae **6**:*770*
Bombinae **4**:*526*
Bombycidae **4**:*494*
Bombycoidea **4**:*494*
bone
  cells **9**:*1168*, 1184
  density **9**:*1168*
  of humans,
    development of
    long bones **9**:*1169*
  marrow *see* marrow,
    bone
  of sirenians **5**:707
  spongy
    in elephants **2**:250,
    251
    in whales **2**:194
bonefish **3**:*436*
bonobo *see* chimpanzee,
    pygmy
Borneo
  pitcher plants **10**:*1303*
  porcupines **7**:900
botfly **4**:539, 547; **8**:*1021*
  sheep, ovoviviparity
    4:549
boutu (Amazonian river
    dolphin) **2**:*195*, 205
Bovidae **10**:*1360*, 1361
bovids
  evolution **10**:1377
  horns **10**:*1364–1365*,
    1369
  skull **10**:1369
Bovinae **10**:*1360*, 1361
bovine spongiform
    encephalopathy
    10:1318
Bovoidea **3**:*356*; **7**:*964*
brachial plexus **5**:587
brachiation **1**:*95*, 96
Brachiosaurus **9**:*1160*
Brachycephalidae **1**:*54*
Brachycera **4**:*538*, 539

Brachylagus idahoensis
    4:*476*
Brachyura **1**:*132*, **5**:*684*
brachyurans **1**:*132*, 133
bracken **2**:*276*, 279, *282*
bracket fungi **6**:771, *773*,
    775, 778
Bradypodidae **3**:*312*;
    9:*1176*
Bradypus **9**:1180
  torquatus **9**:*1176*
  tridactylus **9**:*1176*
  variegatus **9**:*1176*
brain **6**:781, *794*, 795,
    796–799
  blood-brain barrier
    6:795
  imaging the **6**:*799*
Brazil, sloths **9**:*1183*,
    1190, 1193
breathing, voluntary and
    involuntary **4**:520;
    8:1020
breeding aggregates
    3:409
Brentidae **10**:*1344*
Brisingida **9**:*1224*
brittle star **9**:*1224*, 1225,
    1228
bromatia **6**:773
bronchi **4**:566–*567*;
    5:677; **7**:911;
    8:1020; **10**:1374
bronchioles **4**:566–*567*;
    5:677; **7**:911;
    8:1020
brood patch **1**:19
brood pouch **8**:1108,
    1109
Brookesiinae **5**:*624*
Brunner's glands **2**:182
Bryopsis corticulans **8**:1133
"bubble boy disorder"
    (severe combined
    immunodeficiency)
    5:*622*, 623

budding **7**:988
Bufonidae **1**:*54*, 55
Bulbophyllum **6**:832
    globuliforme **6**:828
bulbus arteriosus **8**:1070,
    1104, 1105; **9**:1248
bullfrog **1**:**54–63**
    calls **1**:57, 61, 63
    circulatory and
        respiratory systems
        1:55, *61*
    digestive and excretory
        systems **1**:55, *62*
    external anatomy **1**:55,
        56–57
    nervous system **1**:55,
        60
    reproductive system
        1:55, *63*
    skeletal and muscular
        systems **1**:55, *58–59*
bumblebee **4**:*526*, 527
Bunolagus monticularis
    4:*476*
burrowing
    bivalves **3**:*347*, 351
    hagfish **4**:457
    snakes, skulls of **3**:*403*
bushbuck **10**:*1376*
    skull **10**:*1369*
butterfly **4**:*494*, 495,
    538, 541; **5**:*692*
    antennae **4**:496
    wings **2**:211
butterwort **10**:1303
buzzard
    common, eyes **2**:229
    flight **2**:223
byssal threads **3**:348, 349

C
Cactaceae **8**:*1048*, 1049
Cactoideae **8**:*1048*,
    1049, 1052
cactus **8**:*1048*
    claret cup **8**:*1059*

and drought
    8:1056–1057
external anatomy
    8:1049, *1050–1054*
golden barrel **8**:*1050*
internal anatomy
    8:1049, *1055–1057*
reproductive system
    8:*1058–1059*
saguaro *see* saguaro
    cactus
seeds **8**:1058, 1059
surface area and desert
    survival **8**:1054
caddisfly **4**:*538*
caecilians **1**:*54*; **3**:407;
    6:*800*
caiman **2**:*148*, 149
    dwarf **2**:*148*, 149
        nostrils **2**:*160*
        snout **2**:150
    spectacled **2**:*160*
        calcitonin **2**:268
California
    cacti **8**:*1051*
    chameleons **5**:624
    Calliphoridae **4**:*538*
    Callitrichidae **6**:725
    calls, frog **1**:*55*, 57, 61
    calmodulin **6**:769
    calyx **1**:43
    cambium **1**:*40–41*;
        7:*924*; **8**:*1144*;
        9:*1161*
Cambrian explosion
    9:1157–1158
camel **3**:*356*; **4**:*506*;
    7:*964*; **10**:*1360*
    stomach **10**:1377
camouflage
    caterpillars **4**:499
    chameleons **5**:630
    coelacanths **1**:*123*
    deer **7**:969
    giraffes **3**:359
    hares **4**:*481*

lions **5**:669
lobsters **5**:686
octopuses **6**:815–816
porcupines **7**:902
sea horses **8**:1100
sloths **9**:1183
snakes **3**:400
tigers **5**:668
trout **9**:1282–1283
zebras **10**:1421
see also countershading
Campanulales **7**:*916*
Canada, wolves **10**:1385
cancer **1**:*86*, 87; **10**:1323
Cancer
    antennaris **1**:*132*
    irroratus **1**:*132*
    jordani **1**:*132*
    magister **1**:*132*
    pagurus **1**:*132*
Cancridae **1**:*132*
cancridans **1**:*132*
Candida albicans **5**:*607*
Canidae **3**:*410*; **6**:*844*;
    10:*1382*, 1384
canids, brain
    10:1394–1395
Canis **10**:*1382*, 1384
    lupus familiaris **10**:1388
    rufus **10**:1388
cannon bone **3**:362;
    7:970
Canoidea **5**:*592, 664*;
    7:*928*; **10**:*1324*
capillaries **1**:119–*120*
Capreolus capreolus **7**:*964*
Caprolagus hispidus **4**:*476*
capsid bug **10**:1304
capsomers **10**:*1317*
capybara **7**:*898*
Carabidae **10**:*1344*
carapace **1**:134–135
    of lobsters **5**:686
    of spiders **9**:1254
    of tortoises **9**:*1268*,
        1269, 1272

of turtles **9**:*1197*–1198,
    1200
Carcharhinidae **4**:*460*
Carcharhiniformes
    4:*460*, 461
cardiac ganglion **6**:767
Cardiidae **3**:*342*
Caribbean islands,
    manatees **5**:700
Caribbean Sea, jellyfish
    5:645
caribou (reindeer) **7**:*964*,
    966, *972*, 974
    females with antlers
        7:982
Carnegiea gigantea **8**:*1048*
Carnivora **3**:*410*; **5**:*592*,
    664, 665; **6**:*844*,
    845; **7**:*928*, 929;
    8:*1110*–1111;
    10:*1324*, 1383
carnivores **3**:*410*; **5**:*592*,
    664, 665; **6**:*844*,
    845; **7**:*928*, 929;
    8:*1110*–1111;
    10:*1324*, 1325,
    1382, 1383
    catlike **5**:*592, 664*;
        10:*1324*
    digestive system **8**:1122
    intestines **2**:181;
        8:1122–1123
    land **3**:*410*; **10**:*1382*
    teeth **2**:174; **5**:592
carotenes **8**:1127
carotid arteries **1**:121
carp, scales **9**:*1283*
carpals **9**:*1174*
carpels **7**:*1001*; **8**:1059
carpogonia **8**:1136
carpometacarpus **7**:873
carpospores **8**:*1134*,
    1136
carpus **10**:*1425*
Carretochelyidae
    9:*1194*, *1266*

carrot **1**:*36*
cartilage **9**:1168–1169
  hyaline **9**:*1169*
  in sharks **4**:*464*, 465;
    9:1169, 1242
  in skates and rays
    9:1242
Caryophyllaceae **8**:1049
Caryophyllales **8**:*1048*,
  1049
caryopsis **6**:752
Cassiopea **5**:638, 645
cassowary **6**:*834*, 835
Casuariformes **6**:*834*
cat **5**:*592*, *664*, 665;
  6:*844*; **7**:*928*, 929
  big cats **5**:*664*, 665;
    7:*928*, 932
    eyes **7**:939
  classification **7**:932
  coat patterns **7**:*931*
  Iriomote **7**:932
  Manx **7**:935
  small cats **5**:*664*, *668*;
    7:*928*, 929, 932
    eyes **7**:*939*
    hyoid apparatus
      7:*935*
Catarrhini **1**:*88*; **4**:*550*;
  6:*724*
catecholamines
  2:271–272
caterpillars **4**:504–*505*
  digestive system **2**:*179*
  metamorphosis **2**:267
  moth **4**:*499*, 501, *505*
  nervous system **4**:502
catshark **4**:*460*
  barbeled **4**:*460*
  false **4**:*460*
  finback **4**:*460*
cattle **3**:*356*, 357; **7**:*964*;
  10:*1360*
  stomach **5**:*659*
  wild **10**:*1360*
caudal glands **10**:1389

Caudata **1**:*54*
Caulerpa **8**:1127
Caviidae **7**:*898*
cavy **7**:*898*
Cebidae **6**:725
cebids **6**:*729*, 730
cecal pellets and feces
  4:490–*491*
cecropins **5**:608
cecum **1**:102; **2**:181, 182;
  4:490, 569;
  8:1045; **9**:1220
  of eagles **2**:230
cell biology and genetics
  1:**64–87**
  beginnings of life **1**:65,
    69
  cell constituents
    1:64–65
  cell cycle and cell
    division **1**:*84–87*
  cell external anatomy
    1:*68*
  cell internal anatomy
    1:*69–74*
  see also organelles
cells and their functions
  1:*66–67*
  genes and inheritance
    1:*76–79*
  movement and support
    1:*75*
  nucleic acids **1**:*80–83*
  see also DNA; RNA
cell membrane
  (cytoplasmic
  membrane; plasma
  membrane) **1**:*68*
  bacterial **1**:50, *51*, *52*
cellulose **1**:42
cellulose digestion **6**:739;
  10:1311
  in cows **5**:659
  in deer **7**:979–980
  in even-toed ungulates
    4:506–507

  in giraffes **10**:*1377*
  in hares and rabbits
    4:490–491
  in kangaroos **5**:659,
    660; **10**:*1377*
  in mice **7**:961
  in porcupines
    7:912–913
  in rhinos **8**:1045
  in sirenians **5**:714
  in sloths **9**:*1190*, 1191
  in squirrels **9**:1220
  in wildebeests **10**:1376,
    1377
  in zebras **10**:1435
cell wall **1**:68
  of ciliates **2**:167
  of diatoms (frustule;
    theca) **2**:166, *167*
  of dinoflagellates **2**:167
  of heliozoans **2**:167
  of radiolarians **2**:167
centipede **3**:*332*, 333;
  8:*1074*
  giant *see* giant
    centipede
  house **3**:337, 338
  soil **3**:*332*
  stone **3**:*332*
Central America
  manatees **5**:700
  sloths **9**:*1183*
Centrales **2**:*164*
central pattern
  generators **6**:795
centromeres **1**:76
centrum **9**:1172
Cephalaspidomorphi
  4:*446*, 447
cephalic lobes **9**:1249
cephalium **8**:*1053*, *1056*
Cephalocereus **8**:1052
cephalofoil **4**:462, 470,
  471
Cephalophinae **10**:*1360*,
  1361

Cephalopoda **3**:*342*;
  6:*812*
cephalopods **6**:*812*, 813
  color change **6**:816
cephalothorax *see*
  prosoma
Cephalotus follicularis
  10:*1300*
Ceratodontidae **1**:*122*
Ceratodontiformes **1**:*122*
ceratodonts **1**:*122*
Ceratotherium simum
  8:*1032*
cercariae **7**:987
Cercocebus **6**:*724*
Cercopithecidae **1**:*88*;
  4:*550*; **6**:*724*, 725,
  732
cercopithecines **6**:725,
  739
Cercopithecus **6**:*724*
cerebellum **6**:797
  of chimpanzees **1**:98,
    99
  of dolphins **2**:199
  of elephants **2**:257
  of humans **4**:*564*;
    6:*794*, 795
  of hummingbirds **5**:587
  of manatees **5**:709
  of ostriches **6**:840
  of seals **8**:1118
cerebral cortex **6**:*798*
  of dolphins **2**:199–200
  of elephants **2**:257
  of humans **4**:*564*;
    6:*796*
  of ostriches **6**:840
cerebral ganglion, of
  lobsters **5**:*690*
cerebrospinal fluid **4**:564
cerebrum **6**:797,
  798–799
  of birds **2**:229
    hummingbirds **5**:587
    ostriches **6**:840

of dolphins **2**:199–200
of elephants **2**:257
of humans **4**:*564*;
   6:*794*
of mandrills **6**:737
of pumas **7**:*939*
of whales **3**:385
of wolves **10**:1395
Cereus lamprospermus
   8:1050
Ceriantharia **8**:*1084*
cervix **7**:996
Cervoidea **7**:*964*
Cervus elaphus **7**:*964*
Cetacea **2**:*188*; **3**:*374*;
   4:*506*
cetaceans **2**:189;
   3:374–375
   limb morphology
      2:192, 196
   skin **2**:191
Chagas' disease **5**:622
Chamaeleonidae **5**:*624*
Chamaeleoninae **5**:*624*
chameleon **5**:*624*, 625
   circulatory and
      respiratory
      systems **5**:625,
      632
   color change **5**:*631*;
      6:*789*
   digestive and excretory
      systems **5**:625, *633*
   dwarf **5**:*624*
   external anatomy
      5:625, *626–627*
   Hoehnel's **5**:634–635
   Jackson's **5**:*624–635*;
      7:*997*
   muscular system **5**:625,
      630
   nervous system **5**:625,
      631
   reproductive system
      5:625, *634–635*;
      7:*997*

skeletal system **5**:625,
   628–629
char **9**:1281
cheetah **5**:665, *673*, 678;
   7:932
Cheirostylis **6**:826–827
chelae **1**:135, 139; **5**:687;
   8:1077
chelicerae **9**:1253, 1254
   scorpion **8**:1077
   spider **2**:175; **9**:*1254*,
      1256, 1262
Chelicerata **9**:*1252–1253*
chelicerates **9**:*1252–1253*
chelipeds **5**:687
Chelonia **9**:*1194*, *1266*
chelonians **9**:1267
Cheloniidae **9**:*1194*,
   1266
Chelydridae **9**:*1194*,
   1266
chemoreceptors **6**:790
   housefly **4**:546
   spider **9**:1260
Chenopodiaceae **8**:1049
cherry **1**:37, 38, 43, 44
chickaree **8**:*1148*
Chilopoda **3**:*332*
chimera **4**:470
chimpanzee **1**:**88–105**;
   4:*550*
   ancestral diets **1**:101
   babies **1**:*104–105*
   bipedalism **1**:*94*, 97
   circulatory and
      respiratory systems
      1:89, *100*
   common **1**:89, *90*, 91,
      92, *93*, *95*, *98*, *100*,
      101, *102*, *103*
   digestive and excretory
      systems **1**:89,
      101–102
   external anatomy **1**:89,
      90–91
   facial expressions **1**:*96*

language skills **1**:99
meat eaters **1**:102
muscular system **1**:89,
   95–97
nervous system **1**:89,
   98–99
opposable thumbs and
   toes **1**:*97*
pygmy (bonobo) **1**:89,
   91, *94*
   gestation **1**:103
   promiscuity **4**:573
reproductive system
   1:89, *103–105*
skeletal system **1**:89,
   92–94
social structure **1**:105
strength **1**:97
teeth **4**:569
tool use **1**:97, 104
vocalizations **1**:100
chinchilla **7**:*898*
Chinchillidae **7**:*898*
chipmunk **9**:*1208*
Chironex **5**:640
Chiroptera **3**:*292*
chitin **1**:135; **2**:210–211;
   4:496, 543; **9**:1257
chiton **3**:342
chitrids **6**:770–771
Chlamydiae **1**:*46*, 47
chlorenchyma **8**:*1055*
chloride cells **9**:1290
Chloridoideae **6**:*744*,
   745
chlorocruorin **8**:*1027*
chlorogogen **2**:184, 242
chlorophyll **1**:42, 74
Chlorophyta **2**:*164*;
   8:*1126*, 1127
chloroplasts **1**:42, 74;
   6:*750*, 829–830;
   7:*923*, *924*; **8**:*1145*
cactus **8**:1055
diatom **1**:*74*; **2**:169
sea lettuce **8**:1132

chlorosomes **1**:51
cholecystokinin **2**:272
Choloepus **9**:1179–1180
   didactylus **9**:*1176*
   hoffmanni **9**:*1176*
Cholus cinctus **10**:*1351*
Chondodendron
   tormentosum **6**:761
Chondrichthyes **3**:*436*;
   4:*460*; **8**:*1060*,
   1096; **9**:*1238*, *1280*
chondroblasts **9**:*1169*
chondrocytes **9**:1169
chordae tendineae **1**:*116*
Chordata **2**:*220*; **8**:*1110*
chordates **2**:148; **7**:868
   ancestors, body plan
      2:241
chromatin **1**:79
chromatophores **5**:*631*;
   6:*789*, 816
chromocytes
   9:1283–1284
chromosomes **1**:76;
   4:493; **7**:990
   autosomes **7**:990
   bacterial **1**:50–51
   of fruit flies **4**:543
   and polyploidy **7**:927
   replication **1**:*84*
   sex (X and Y) **1**:*76*, 78;
      7:990, 991
chronic wasting disease
   (CWD) **7**:975
chrysalises **4**:495
   butterfly **4**:504
chrysolaminarin **2**:169
chyme **2**:181–182, 204
Chytridiomycota
   6:*770*
cilia **1**:75
ciliary cone apparatuses
   8:1090
ciliates **2**:*164*, 165;
   8:*1126*
   cell wall **2**:167

Ciliophora **1**:*20*, 21;
  2:*164*; **8**:*1126*
cinclides **8**:1089
Cingulata **3**:*312*
circulatory system
  1:**106–121**
  open and closed
      systems **1**:107
  regulation and control
      1:*121*
  see also arteries; blood;
      capillaries; heart;
      veins
circumesophageal
      connectives **2**:215
civet **5**:*592*, *664*; **7**:*928*;
  10:*1324*
cladodes **8**:1051
clam
  catch muscles **6**:768
  dipper **3**:*342*, 353
  giant see giant clam
  nut **3**:352
  razor **3**:*342*, 343, 351
  soft-shell **3**:*342*
  vent **3**:*342*, 343, 352
clavicle (collarbone)
      9:1174
  of chimpanzees
      1:*93*
  of monotremes
      7:889
  of pumas **7**:937
claws
  extendable, of lions
      5:*671*
  pelvic **3**:*400*
  retractable, of pumas
      7:*931*
Cleistes **6**:826
Clever Hans **10**:1431
Clitellata **2**:*234*
clitellate worms **2**:*234*,
  235
clitellum **2**:235
clitoris **7**:996

cloaca **3**:326, 330
  of amphibians
      frogs **1**:63
      newts **6**:*810*
  of birds **5**:589
      albatrosses **1**:18, 19
  of mammals **2**:185
      marsupials **2**:*185*
      monotremes **2**:185;
          7:883, *894*,
          895–896
  of reptiles
      crocodiles **2**:161, 162
      turtles **9**:1205
  of snakes **3**:408
cloacal bursae **9**:1275
clonal selection **5**:620
clotting **1**:*119*
clown fish, changing sex
      7:*998*
club moss **2**:*276*, 284
clubtail **2**:*208*
cnidae see nematocysts
Cnidaria **5**:*636*; **8**:*1084*,
  1085, 1087
cnidarians **5**:*636*, 637;
  8:*1084*, 1085, 1087
  body wall **5**:*642*
  nerve cells **8**:1090
  venoms used by **5**:640
cnidocytes **5**:637, *640*;
  8:1086, 1092
cobra, fangs **3**:*404*
Coccinellidae **10**:*1344*,
  1345
coccyx **1**:*93*; **4**:558;
  9:1173
cochlea **10**:1413
cockle **3**:*342*, 343, 345,
  347–348, 350
Codiaeum variegatum
  10:*1303*
codons **1**:81
coelacanth **1**:**122–131**
  African **1**:*122*, 124,
      125, 127, 129, 130

buoyancy **1**:129
circulatory and
      respiratory systems
      1:123, *129*
  digestive and excretory
      systems **1**:123, *130*
  external anatomy
      1:123, *124–125*
  Indonesian **1**:*122*, 124
  muscular system **1**:123,
      127
  nervous system **1**:123,
      128
  reproductive system
      1:123, *131*
  scales **9**:1283
  skeletal system **1**:123,
      126
  swimming **1**:*125*, 127
Coelacanthiformes **1**:*122*
coelenteron **5**:637, 642,
      644–*645*; **8**:*1088*,
      1089, 1091, 1093;
      9:*1166*
Coelodonta **8**:*1036*
coelom, of earthworms
      2:234, 239;
      9:*1163*
coevolution see
      evolution,
      convergent
Coleoidea **6**:*812*, 813
coleoids **6**:*812*, 813
Coleoptera **4**:*494*, *538*;
      5:*692*; **10**:*1344*,
      1345
collagen **9**:1184
collenchyma cells **1**:67;
      7:924
Colletidae **4**:*526*
colobines **6**:725, 739
colobus see monkey,
      colobus
Colombia, porcupines
      7:903
colon **2**:182

color change
  chameleons **5**:*631*;
      6:*789*
  octopuses, squid, and
      cuttlefish
      6:815–*816*
  sailfish **8**:1064
colostrum **10**:1380
columella **1**:60; **6**:*838*;
      9:1170
Commelinidae **6**:825
communication
  crocodile **2**:152
  elephant **2**:*253*, 257,
      262, 263
  giraffe **3**:366
  honeybee **4**:534
  hyena **5**:600–*601*
  kangaroo **5**:657
  lions **5**:669, 676
  whale **2**:201
  wolf **10**:1388, *1389*,
      1401–1402
  zebra **10**:*1420*
  see also calls; scent
Comphotherium **2**:247
complement cascade
      5:609, 611, 614
Concentricycloidea
      9:*1224*
conceptacles **8**:*1136*,
      1137
conch, queen **9**:*1164*
cones
  giant sequoia **8**:*1140*,
      1143, 1146–1147
  larch **8**:*1143*
  pine tree **7**:*1000*, 1003;
      8:*1146*, 1147
Congo, Democratic
      Republic, mystery
      chimpanzee **1**:91
conidia **6**:777
conifer **8**:*1138*–1139
  external anatomy
      8:*1140–1143*

internal anatomy
    8:*1144–1145*
reproductive system
    8:*1146–1149*
Coniferophyta 8:*1138*
Coniophora 6:*779*
conjugation 1:24, *25*, 53
Connochaetes
    gnou 10:*1360*
    taurinus 10:*1360*
conus arteriosus 4:456;
    9:1248
copepod, filter feeding
    2:175
Copiapoa 8:1054, 1055
    conglomerata 8:*1050*
Coprinaceae 6:*770*
coprodaeum 2:162
coprophagy 2:*182*;
    4:*491*; 7:961
coracoid bones 5:585;
    7:*873*, 888–889;
    9:1173, 1174, 1200
coral 8:1085, 1087
    black 8:*1084*
    crown-of-thorns
        starfish and 9:1235
    soft 8:1087
    stony 8:*1084*, 1087
Corallimorpharia 8:*1084*
    coral reefs 8:1087;
        9:1167
    giant clams 3:*343, 344*
cordgrass 6:*744*, 745
    common 6:*744*
    small 6:*744*
    smooth *see* marsh grass
corms 6:827, *828*
corn (maize) 6:*749*, 753
cornea 4:563
    flat 7:876
corolla 1:43
coronary arteries 1:115
Coronatae 5:*636*
Coronaviridae 10:1314
Coronavirus 10:1314

corpora allatum 2:215
corpora cardiaca 2:215
corpus callosum 5:657
corpus luteum 7:*995*,
    999
Cortaderia selloana 6:751
cortex, of green plants
    1:*65*; 8:*1055*;
    10:*1306*
cortisol (hydrocortisone)
    2:272
Coryanthes 6:833
Corybas 6:832
cotton 6:752
cottontail 4:*476*
countercurrent heat
        exchangers 1:120
    in eagles 2:231
    in fish 8:1069
    in otters 6:856
    in penguins 7:*877–878*
    in seals 8:1124
    in whales 2:202; 3:391
countershading
    sailfish 8:*1063*
    whales 3:379
courtship
    crab 1:135, 140, *141*
    crocodile 2:162–163
    hippopotamus 4:519
    lobster 5:691
    newt 6:*810*–811
    octopus 6:823
    porcupine 7:914
    rhino 8:1047
    scorpion 8:*1082*
    sea horse 8:1109
    spider 9:1264, 1265
    stingray 9:1251
    turtles 9:1206
coxal glands 8:1081;
    9:1263
coyote 10:*1382*
crab 1:*132–141*; 5:*684*
    Atlantic rock 1:*132,
        133*, 141

blue 1:136
boxer 1:135
circulatory and
    respiratory systems
    1:133, *138*
coral 1:*141*
digestive and excretory
    systems 1:133, *139*
Dungeness 1:*132*, 133
edible 1:*132, 134, 136,
    137, 138, 139, 140*
external anatomy
    1:133, *134–135*
fiddler 1:135, 140
green 1:*136*
hairy rock 1:*132*
hermit 1:*133*, 135;
    8:1087
horseshoe 8:1017;
    9:*1252*
immunity 5:611
land 1:*132*
mole 1:133
molting 1:141
muscular system 1:133,
    136; 6:762, 763
nervous system 1:133,
    137
Pacific rock 1:*132*
pea 3:347
reproductive system
    1:133, *140–141*
robber 1:136
sand 1:133
spider 1:*132*
swimming crabs 1:*132*
true 1:133
walking sideways 1:*136*
cranium 7:938
crassulacean acid
    metabolism (CAM)
    6:830; 8:*1030,
    1057*
Crataegus 1:*36*, 37
crayfish 1:133
    claws 2:175

muscles 6:762
    heart 6:767
Crenarchaeota 1:*46*
Creutzfeldt-Jakob disease
    (CJD) 10:*1318*
cribriform plate 3:305
Cricetinae 7:*946*
cricket
    excretion 2:*187*
    sound detection 6:790
Crinoidea 9:*1224*, 1225
crinoids 9:*1224*
crocodile 2:**148–163**
    African dwarf 2:149
    American 2:*148*, 149,
        150
    blood and air flow
        2:*159*
    brain and spinal
        cord 2:*158*
    digestive system
        2:*161*
    jaws 2:*153*
    muscles 2:*156*
    palatal and nasal
        valves 2:*160*
    skeleton 2:*154*
    ancient 2:155
    biggest ever 2:154
    biting and swallowing
        2:156–157, 161
    brain 2:158
    breathing 8:1028
    breathing underwater
        2:160
    circulatory system
        1:*112*; 2:149,
        159–160
    communicating by
        smell 2:152
    differs from alligators
        2:151
    digestive and excretory
        systems 2:149, 161
    dwarf 2:148
    ears 2:151

crocodile *(continued)*
  eggs **2**:162, 163
  external anatomy
      2:149, 150–152
  eyes **2**:150–151, 158
  jaws **2**:151, 153, 158
  Johnston's **2**:156
  moving and swimming
      2:150, 156, 157
  muscular system **2**:149,
      156–157
  nervous system **2**:149
  Nile **2**:*157*
  nostrils **2**:150, 151
  reproductive system
      2:149, 162–163
  respiratory system
      2:149, 160
  ripple detectors
      2:158
  salt glands **2**:*152*, 161
  skeletal system **2**:149,
      153–155
  skull **2**:153
  tail **2**:157
  teeth **2**:151, 153
  true **2**:148, 149
crocodilians **2**:149;
    5:*624*; **9**:*1194*,
    1266
  scales **2**:151–152
  snouts **2**:150
Crocodylia **2**:*148*; **5**:*624*;
    9:*1194, 1266*
Crocodylidae **2**:*148*
Crocodylus **2**:*148*
Crocuta crocuta **5**:*592*
crop
  of birds **2**:178
    eagles **2**:230
  of dragonflies **2**:214
  of hummingbirds **5**:589
crop milk **7**:879
Crossaster papposus
    9:*1224*
crossbreeds, deer **7**:981

Crustacea **1**:*132–133*;
    5:*684*; **8**:*1074*;
    9:*1252*
crustaceans **1**:*132–133*;
    3:332; **5**:*684*–685;
    8:*1074*; **9**:*1252*
  digestive system **2**:179
  excretory system **2**:187
  exoskeleton **9**:1163
  hormones **2**:269
crust fungus **6**:*779*
Cryptobranchidae **6**:801
Cryptodira **9**:*1194,
    1266*, 1273
ctenidia **6**:*821*; **8**:1017
Ctenophora **5**:636–637
ctenophores **5**:636–637
Cubozoa **5**:*636*, 638;
    8:*1084*
cubozoans **5**:638, 639
cud, chewing
    10:1375–1376
cup fungus **6**:*774*, 775
Cupiennius **9**:1265
Cupressaceae **8**:*1138*,
    1139
curare **6**:761
Curculionidae **10**:*1344*,
    1345
cuticle **9**:1258
  of dragonflies **2**:210
  of lice **5**:695
  of lobsters **5**:687;
      9:*1167*
  of spiders **9**:1257
cuttlefish **6**:*812*, 813
  color change **6**:*816*
  ink **6**:817
  large nerve cells **6**:820
cyamid (whale louse)
    3:*378*
cyanobacteria (blue-
      green algae) **1**:*46*,
      47, 48, *49*; **8**:1015
  on sloths' fur **9**:1178,
      1182, 1183

Cyathea **2**:280
cycad **8**:*1138*
Cycadophyta **8**:*1138*
cyclic adenosine
      monophosphate
      (cAMP) **2**:265
cyclins **1**:85
Cyclopes spp. **3**:*312*
Cyclorrhapha **4**:*538*, 539
Cydonia **1**:*36*
Cyemetoidei **3**:*436*, 437
Cylindropuntia **8**:1049
Cynomys **9**:1208
Cyperaceae **6**:*744*
Cyprepedioideae **6**:*824*
cypress **8**:*1138*
  California **8**:1142
  Montezuma **8**:1139
  swamp **8**:1143
Cypripedium **6**:832
cystic fibrosis **1**:79
cytoplasm **1**:73
cytoplasmic membrane
      see cell membrane
cytoskeleton **1**:*73, 75*
  diatom **2**:169–170
cytosol **1**:50

D

Dactylaria brochopaga
    6:774
Dama dama **7**:*964*
damselfly **2**:*208*, 209;
    10:*1312*
  nymphs, breathing
      underwater **2**:216;
      8:1017
dandelion, seeds **7**:*988*
darner **2**:*208*, 209
Darwin, Charles **10**:1309
Dasypodidae **3**:*312*
dead man's fingers
    8:*1133*
death cap **6**:771
Decapoda **1**:*132*, 133;
    5:*684*, 685

Decapodiformes **6**:*812*
decapods **1**:*132*, 133;
    5:*684*, 685; **6**:*812*
  claw muscles **6**:762
  neurogenic hearts
      6:767
decompression sickness
      see bends, the
deep-sea environment *see*
      ocean depths
deer **3**:*356*, 357; **7**:*964*,
      966, 967, 968–969;
      10:*1360*
  antlers **7**:*965*, 966, 969,
      971–*972*, *974*, 975,
      980, *982*
  brocket **7**:981
  Chinese water **7**:969,
      983
  chronic wasting disease
      (CWD) **7**:975
  crossbreeds **7**:981
  "endurance hunting"
      of **7**:974
  fallow **7**:*964*, *966*, *974*,
      983
  lesser Malay mouse
      7:*969*
  mouse **3**:*356*
  mule **7**:*964*, 982
  muntjac **7**:*964*, 969,
      981
  musculature **7**:973–974
  musk **3**:*356*; **7**:*964*;
      10:*1360*
  names for males,
      females and young
      7:983
  pampas **7**:*969*
  red *see* red deer
  red brocket **7**:*969*
  roe **7**:*964*, 966, 982
  sika **7**:981
  skeleton **7**:*970*
  with white rumps
      7:967

white-tailed **7**:*964, 966, 980,* 982
defecation **2**:186
defense
   giraffe **3**:360
   millipede **3**:*340*
   sea anemone **8**:1090
   tarantula **9**:1256
   tortoise **9**:*1268*
Delphinidae **2**:*188*
deltoid muscles **1**:15
deltoid processes **1**:96
dementia, senile **6**:*799*
dendrites **6**:782
dendritic cells **5**:611
Dendrobium cucumerium **6**:828
Dendrobranchiata **5**:*684*
dendrochronology **8**:1145
Dendrolagus **5**:649
Dennstaedtiaceae **2**:*276*
denticles, of sharks **4**:*463*
deoxyribonucleic acid *see* DNA
depressor caudae muscle **2**:228
Dermaptera **3**:297
deserts, antelope jackrabbits **4**:479
Desmatosuchus **2**:*155*
desmosomes **6**:766
destroying angel **6**:771
deutocerebrum **3**:337
dewclaws **7**:930; **10**:*1389*
diabetes **4**:*571*; **5**:622
diaphragm **3**:329; **4**:567; **5**:677; **7**:911, 958–959; **8**:*1028*; **9**:1219; **10**:1432
diaphragmaticus muscle, crocodile **2**:160; **8**:1028
diastema **3**:*427*; **5**:655; **7**:905, *953*; **9**:1215
diastole **1**:117

diatom **1**:*74*; **2**:**164–171**
   centric **2**:*164,* 165, 168
   colonial **2**:*170*
   and diatomaceus earths **2**:166
   differ from bacteria **2**:168
   external anatomy **2**:165, *166–167*
   fossils **2**:169
   internal anatomy **2**:165, *168–170*; **9**:*1157*
   pennate **2**:*164,* 165
      gliding movement **2**:168
   reproductive systems **2**:165, *171*; **7**:*987*
   shapes **2**:*167*
Dicamptodontidae **6**:801
Dicerorhinus sumatrensis **8**:*1032*
Diceros bicornis **8**:*1032*
Dicksonia **2**:280
dicots **1**:*36,* 37; **6**:*744, 824*; **7**:*916,* 917; **8**:*1048*; **10**:*1300*–1301
Dicotyledonae **1**:*36*; **6**:*744, 824*; **7**:*916*; **8**:*1048*
dictyostele **2**:282
diencephalon **10**:1395
digestive and excretory systems **2**:**172–187**
   invertebrate excretory systems **2**:187
   intestines **2**:183
   mouthparts **2**:175
   vertebrate excretory systems **2**:186–187
   intestines **2**:180–183; **7**:*943*
   liver, pancreas, and gallbladder **2**:184

lymphoid tissue **2**:181
   mammalian **6**:*741*
    stomach **2**:177–178
    teeth **2**:173–174
digestive glands **2**:*172*
   fundic glands **2**:*177*
digitigrade stance **3**:418
   deer **7**:970
   elephants **2**:253
   Eurasian otters **6**:852
   hares **4**:*483*
   hyenas **5**:596
   lions **5**:670
   monkeys and apes **6**:733
   wolves **10**:1390
dik-dik, territory-marking **2**:274
Dilleniidae **10**:*1300,* 1301
dimethyl sulfide **1**:16
dimorphism, sexual
   in humans **4**:573
    the pelvis **4**:*558*
   in lions **5**:665
   in mandrills **6**:726
   in snakes **3**:400
   in spiders **9**:1264
   in woodpeckers **10**:1407
dingo **7**:885; **10**:*1382*
dinoflagellates **2**:*164,* 165; **8**:*1126*
   cell wall **2**:167
Dinophyta **2**:*164*
Dinornithiformes **6**:*834*
dinosaurs, skeleton **9**:1160
dioecious animals, centipedes **3**:336
dioecious plants **7**:1002
Diomedea **1**:11
   albatrus **1**:*10*
   chlororhynchos **1**:*10*
   chrysostoma **1**:*10*

   epomophora **1**:*10*
   exulans **1**:*10*
   immutabilis **1**:*10*
Diomedeidae **1**:*10*
Dionaea **10**:1301
   muscipula **10**:*1300*
Diplodocus **9**:1160
diploid cells **1**:76, 87; **7**:*985,* 990
Diplopoda **3**:*332*
Dipodidae **7**:*946*
Diprotodontia **5**:*648, 655*
Diptera **4**:*538,* 539; **5**:*692*
dipterans **4**:539
Discomycetes **6**:*770*
diving
   manatees **5**:712
   platypuses **7**:893
   seals **8**:1117, 1121
   swordfish **8**:1064
   whales **2**:202, 203
diving petrel **1**:*10,* 11
DNA **1**:65, 79, *80–81*; **7**:*984*
   bacterial **1**:50–*51,* 53
   kinetoplastid **1**:24
   mitochondrial (mDNA) **1**:83; **8**:1146
   mutation **1**:85
   replication **1**:*84*–87
   vaccines **1**:83
   see also plasmids
dog **3**:*410*; **5**:*592,* 664; **6**:*844*; **7**:*928*; **10**:*1324, 1382,* 1384
   detecting seizures **10**:1395
   domestic **10**:*1388*
   sniffer **10**:1396
   tail-wagging **10**:1393
   true **10**:*1382,* 1384
dogfish, egg sacs **7**:*1004*

dolphin **2:188–207**; **4:**_506_
  beak or snout **2:**192, 195
  blowhole **2:**195, _196, 203_
  blubber **2:**191, 198
  bottle-nosed **2:**_190_
    brain **2:**201
    calves **2:**207
    friendship alliances **2:**206
    hunting by **2:**205
    neck **2:**195–196
  brain **2:**201
  circulatory and respiratory systems **2:**189, 202–203
  color **2:**193
  common **2:**_188_, 190, 193
    vertebrae **2:**195
  digestive and excretory systems **2:**189, 204–205
  DNA collection from **2:**206
  echolocation **2:**193, _200_, 201
  external anatomy **2:**189, 191–193
  eyes **2:**193
  flippers **2:**197
  hunting by **2:**200, 205
  intelligence **2:**201
  long-beaked common **2:**_188_, 193
  muscular system **2:**189, 197–198
  nervous system **2:**189, 199–201
  oceanic **2:**_188_, 190
  play **2:**_207_
  reproductive system **2:**189, 193, 206–207
  respiratory system **2:**194
  Risso's **2:**204
  river **2:**_188_, 190, 193, 205
    skull **2:**_195_
  senses **2:**200–201
  short-beaked common **2:**_188, 191, 194, 198_
    circulatory system **2:**_202_
    digestive and excretory systems **2:**_204_
    musculature **2:**_197_
    nervous system **2:**_199_
    reproductive systems **2:**_206_
    skull **2:**_196_
  skeletal system **2:**189, 194–196
  skin **2:**191, _192_
  sounds **2:**200
  spotted **2:**_200_; **8:**_1031_
  swimming **2:**192, 197, 198
  tail **2:**197
  taste **2:**200
  teeth **2:**189, _195_, 204
domestic animals
  brain size **10:**1394
  dogs **10:**_1388_
  horses **10:**1423
Donacidae **3:**_342_
dopamine **4:**565
dorsoventral muscles **6:**_762_
douroucouli **6:**_729_
dragonfly **2:208–219**; **4:**528; **5:**_692_
  circulatory and respiratory system **2:**209, 213, 216–217

digestive system **2:**214
emperor **2:**_208_, 209
  circulatory system **2:**_216_
  external anatomy **2:**_210_
  internal anatomy **2:**_214_
  nervous system **2:**_215_
  reproduction **7:**_996_
external anatomy **2:**209, 210–213
eyes and vision **2:**212, 215
flight **2:**212, 214
hunting by **2:**212–213
internal anatomy **2:**209, 214
molting **2:**_211_, 213
muscles **2:**214
nervous system **2:**209, 215
nymphs **2:**218–219
  anatomy **2:**213
  breathing underwater **2:**216, _217_; **8:**1017
  mouthparts **2:**219
  movement **2:**216, _217_
rectum **2:**183, 216
reproductive system **2:**209, 213, 218–219
wings **2:**211–212, 214
Drosera **10:**_1300_, 1301
Droseraceae **10:**1301
Drosophila melanogaster **4:**_543_
Drosophyllum **10:**1301
duck **5:**695
Dufour's gland **1:**30; **4:**532–533
dugong **2:**248; **5:**_700_, 701, 702–703
  birth **5:**717

  food and digestive system **5:**713, _714_
  mating **5:**716
  senses **5:**710
  skeleton **2:**250; **5:**_706_
  skull **5:**_705_
  snout **5:**_704_
  tail **5:**_703_
Dugong dugon **5:**_700_
Dugongidae **5:**_700_
duiker **10:**_1360_, 1361
dung fly **4:**548
duodenum **2:**181, 182, 204; **4:**568, _569_; **10:**1399
dwarfism **1:**79

E

eagle **2:220–233**
  ancestors **2:**223
  bald **2:**_220_, 221, _222_
    brain and spine **2:**_229_
    guts and kidneys **2:**_230_
    muscles **2:**_227_
    respiratory system **8:**_1019_
    skeleton **2:**_225_
    skull **9:**_1170_
    toes **2:**224
    wings **3:**_299_
  bill **2:**223–224, 226, 228, 230
  booted (true eagles) **2:**_220_, 221
  circulatory and respiratory systems **2:**221, 231
  digestive and excretory systems **2:**221, 230
  eggs **2:**232–233
  external anatomy **2:**221, 222–224
  eyes **2:**224, 229
  feathers **2:**222, 223, 228

fish **2**:*220*, 221
    toes **2**:224
flight **2**:222, 227, 231
Haast's **2**:223
harpy **2**:*220*, 221;
    7:*902*
legs and feet **2**:224, 226
life expectancy **2**:*233*
muscular system **2**:221,
    227–228
neck **2**:224, 226, 228
nervous system **2**:221,
    229
regurgitation by **2**:230
reproductive system
    2:221, 232–233
skeletal system **2**:221,
    225–226
snake **2**:*220*, 221, 224
tail **2**:224, 225
temperature regulation
    2:231
wings **2**:222, 223
    bones **2**:*226*
    muscles **2**:227–228
ear, human **4**:*565*; **6**:*789*
earth star **6**:771
earthworm **2**:**234–243**
brain **2**:240
breaking in two and
    regeneration **2**:237
circulatory and
    respiratory systems
    1:107; **2**:235, 241
digestive system **2**:183,
    184, 235, 238, 242
excretory system
    2:235, 237, 238,
    242
external anatomy
    2:235, 236–237
giant **2**:*234*
internal anatomy
    2:235, 238–239
movement **2**:237, 239;
    6:762

muscles **2**:238, 239;
    6:762
nervous system **2**:235,
    240; **6**:*792–793*
reproductive system
    2:235, 238, 243
response to attack
    2:240
sense organs **2**:240
skeletal system **2**:239;
    9:*1163*–1164
skin **2**:236, *237*
ecdysone **2**:267
echidna (spiny anteater)
    7:*882*, 888
brain **7**:*890*
long-beaked **7**:*882*,
    883
long snout of **7**:*886*,
    888, 889
penis **7**:*895*
pouch **7**:*897*
remnants of spurs
    7:887
short-beaked **7**:*882*,
    883, *886*
skeletal system
    7:888–889
Echinocactus platyacanthus
    8:1050
Echinodermata
    9:*1224*–1225
echinoderms
    9:*1224*–1225
body shape **9**:1228
gas exchange by
    8:*1012*
nervous systems
    6:793
related to chordates?
    9:1237
skeletal systems **9**:1164
Echinoidea **9**:*1224*, 1225
echolocation
by fruit bats **3**:305
by whales **3**:375, 388

whales and dolphins
    2:189, 193, *200*,
    201
E. coli see Escherichia coli
ectoderm **7**:1004
ectoplasm, of amoebas
    1:23
Edentata **9**:1185
edentates (xenarthrans)
    3:313, 321
eel **4**:*436*
boatail snipe (arrow)
    4:*436*
deep-sea **4**:*436*
gulper *see* gulper eel
    gills **4**:*443*
pelican **4**:*436*, 438, 444
head pores **4**:*442*
single-jaw **4**:*436*
larvae **4**:445
swallower **4**:*436*, 437
eggs
bird **10**:1417
    albatross **1**:19
    ostrich **6**:*843*
    incubation **5**:591
reptile **5**:635
    alligator or crocodile
        2:162, *163*
    python **3**:405
    tortoise **9**:*1278–1279*
    turtle **9**:1278
see also ova
Egypt, wolves **10**:1385
Eira **10**:*1324*
eland **10**:1361, *1364*,
    1376
Elasmobranchii **9**:*1238*,
    1239
Elasmotherium **8**:*1036*
electric organs, in
    electric rays **9**:*1246*
electron transport chain
    1:*72*
electroplaques **9**:1246
electroreceptors **6**:790

in coelacanths **1**:123,
    124–125
in echidnas **7**:892
in platypuses **6**:*790*;
    7:*891–892*
in sharks and rays
    4:470; **6**:*787*;
    9:*1247*
elephant **2**:174, **244–263**
African **2**:*244*, 246
    body shape **2**:246
    feet **2**:249
    skull **2**:251
    trunk **2**:*246*, 247
    tusks **2**:247
African forest **2**:*244*
    communication
        2:257
    size **2**:246
African savanna **2**:*244*
    circulatory system
        and blood **2**:258
    ears **2**:248
    feet **2**:*252*
    muscular system
        2:*254*
    in musth **2**:263
    nervous system
        2:*256–257*
    reproductive system
        2:*262*
    size and weight **2**:246
    skull **9**:*1171*
    teeth **2**:*261*
    trunk and head
        2:247, 255, *257*
    tusks **2**:247, *251*
    vertebrae **2**:251
ancient **2**:247
Asian **2**:*244*, 245
    body shape **2**:246
    feet **2**:249
    food intake **2**:260
    skull **2**:251; **9**:*1171*
    teeth **2**:*261*
    trunk and head **2**:*247*

elephant *(continued)*
  tusks **2**:*251*
  vertebrae **2**:251
  weight **2**:246
brain **2**:256–257
circulatory and
    respiratory systems
    2:245, 258–259
communication **2**:*253*,
    257, 262, 263
digestive system **2**:245,
    260–261
dung **2**:261
ears **2**:248
  flapping **2**:*254*, 255
energy efficiency
    2:259
external anatomy
    2:245, 246–249
extinct **4**:511
feet **2**:*252*, 253
heart rate **2**:259
lymphatic system **2**:259
meeting mates **2**:262
moving **2**:252–253
muscular system **2**:245,
    254–255
in musth **2**:262, 263
nervous system **2**:245,
    256–257
reproductive system
    2:245, 262–263
sense organs **2**:257
size **2**:246, 249
skeletal system **2**:245,
    250–253
skin **2**:248
teeth **2**:174, 260–261
tool use **2**:257
trunk **2**:*246*, 247, *249*
  muscles **2**:255
tusks **2**:174, 247–248,
    251
water intake **2**:260
elephant bird **6**:837
elephant fish **4**:*470*

Elephas **2**:*244*
  maximus **2**:*244*
Elopiformes **3**:*436*
Elopomorpha **3**:*436*
elytra **2**:211; **4**:542
  of beetles
    10:1346–1347,
    1356
embryo, in animals
    7:*1004–1005*
  bone and skull
    development in
    vertebrates **9**:*1169*,
    1170–1171
emu **6**:*834*
Enaliarctos mealsi **8**:*1113*
encephalization quotient
    (EQ) **4**:564
endocrine and exocrine
    systems **2**:**264–275**;
    6:791
endocrine system
    2:264–265
  vertebrate **2**:265,
    270–273
  see also hormones
exocrine system **2**:264,
    265, 274–275
  skin glands **2**:*265*;
    9:1218
  endocuticle **2**:210, 213
endoderm **7**:1005
endodermal cells **8**:1145
endodermis **6**:831
endometrium **7**:999
endoplasm, of amoebas
    1:23
endoplasmic reticulum
    1:*69*, 70–*71*
  rough **1**:70–71
    of diatoms **2**:169
  smooth **1**:71
Endopterygota **1**:*26*
endopterygotes **1**:*26*;
    4:495, 526,
    538–539

endosperm **6**:752–753
endosterna **9**:1258
endosternite **9**:1258
endosymbiont theory
    1:69
Enhydra lutris **6**:*844*
Entamoeba histolytica **1**:21
Enteromorpha **8**:1127,
    1131
Enterovirus **10**:1314
enzymes, digestive
    10:1339–1340
eosinophils **1**:*108*, 109;
    5:611
epaxial muscles **6**:853
ephyra **5**:647, *648*
epicingulum **2**:*167*
epicuticle
  dragonfly **2**:211, 213
  lobster **5**:687
Epidendroideae **6**:*824*
epididymis **4**:*572*; **7**:995
epiglottis **2**:176
epinephrine **1**:121;
    2:271–272
epiphytes
  cacti as **8**:1050
  orchids **6**:825, 826,
    827, 830, *831*
epipubic bones **5**:654,
    655; **7**:889
epithelial cells **1**:67
Epostoa **8**:1053
Eptatretinae **4**:448
Eptatretus **4**:*446*, 448,
    451, 455, 458
Equidae **8**:*1032*; **10**:*1418*
equids **10**:1419
  brain **10**:*1430*–1431
Equus **10**:*1418*
  caballus **10**:1419, 1423
erector pili muscles
    7:937
erector spinae muscles
    4:561
Erethizontidae **7**:*898*

endosperm **6**:752–753
Ericales **10**:1301
ermine (stoat; short-
    tailed weasel)
    10:1326–1327,
    1328, *1329*, *1331*,
    1342
  reproduction **10**:1343
  size **10**:1340
  skull **10**:*1333*
  strength **10**:1334
Erycinidae **3**:*396*
erythrocytes *see* blood
    cells, red
erythropoietin **1**:108
Escherichia coli (*E. coli*)
    1:*47*, 48, *50*, *51*, *80*
  viruses attacking
    10:*1320*
Eschrichtidae **3**:*374*
Eschrichtius robustus **3**:*374*
Esociformes **9**:*1280*
esophagus **2**:176
estivation, amphibians
    8:1030
estrogen cycle **2**:*269*
estrogens **7**:998–999
estrus cycle, of kangaroos
    5:663
ethylene, a plant
    hormone **2**:266
Eubacteria **1**:*46*
Eubalaena
  australis **3**:*374*
  glacialis **3**:*374*
Eudyptes **7**:*868*
Eudyptula **7**:*868*
eukaryotes **1**:69; **2**:170
Eunectes
  beniensis **3**:*396*
  deschauenseei **3**:*396*
  murinus **3**:*396*
  notaeus **3**:*396*
Eupharyngidae **3**:*436*
Eupholis **10**:*1358*
euphorbia **8**:1052,
    1054

Europe
ground squirrels
9:*1212*
wolverine 10:*1330*
Eurotamandua 3:315
Eurotrochilus inexpectatus
5:585
Euryarchaeota 1:*46*
Eurypelma 9:1258, *1261*
Eurypharynx pelecanoides
3:437, 438
eurypterids 8:1077
Eutheria 1:*88*; 2:*188,
244*; 3:*374*; 4:*506*;
5:*648, 700*; 6:*724,
844*; 8:*1032*;
10:*1418*
evolution,
convergent
ant-eating mammals
3:315
arthropod eyes 3:337
beak of echidnas and
birds 7:889
between octopuses
and vertebrates
6:820
bill of platypuses and
birds 7:889
deep-sea fish 3:440
euphorbias and cacti
8:1052, *1054*
hawkmoth ears 4:498
Old World and New
World porcupines
7:901
orchids and insects
6:832
and swimming
underwater 3:379
of lice 5:695
"evolutionary arms race"
5:608
excretory systems *see*
digestive and
excretory systems

exocuticle 2:210, 213;
9:*1258*
exoskeleton
of arthropods 9:1165
centipedes 3:334
dragonflies
2:210–211
honeybees 4:529–530
houseflies 4:542–543
lice 5:695
scorpions
8:1076–1077
spiders 9:1257
of crustaceans 9:1163
crabs 1:*134*–135
lobsters 5:*686*;
9:*1167*
expansor secondarium
muscles 2:227–228
extensor carpi radialis
1:*15*
extensor carpi ulnaris
1:*15*
extracellular matrix
(ECM) 1:68
extracleithrum 1:126
eyes
cat 5:*668*, 676; 7:*939*
cephalopod 6:788
compound 1:*137*;
4:*496*, 502, *534,
539*
hagfish 4:*453*
human 4:*563*; 6:*788*
lobsters 5:686–687
muscles 6:*765*
octopus 6:*788*, 817,
820
shark 4:469–470, *471*
simple
of giant clams 3:*353*
of scorpions 8:1079
snake 3:*399*, 400, 406
superposition 4:502
third eye *see* pineal
gland

vertebrate 6:788
whale 3:*387*
see also ocelli
eyespots, of starfish
9:1232

F

Fabales 1:*36*, 37
falcon 2:220
African pygmy, feet
2:*226*
Falconidae 2:220
Falconiformes 2:220
fallopian tubes 4:*572*
fan worm 8:*1027*
fascicles 5:697; 6:763,
764
fasciculus aberrans 5:657
fat body 10:1353
feathers 1:13; 2:224
albatross 1:13
contour 1:13; 2:*224*
eagle 2:222, 223
hummingbird 5:*584*
penguin 7:871–872,
875, 881
woodpecker 10:*1406,*
1407, 1409
feather star 9:1225, 1228
feces 2:185–186
elephant dung 2:261
Felidae 5:*592*; 6:*844*
Feliformia 10:*1324*
Felis 5:*664*; 7:*928*, 929
concolor 7:*928*
lynx 7:*928*
pardalis 7:*928*
silvestris 7:*928*
Feloidea 5:*592, 664*;
7:*928*; 10:*1324*
femur 9:*1174*
seal 8:1116
fern 2:276–**285**
adder's-tongue 2:*276,
280*
energy store 2:282

eusporangiate 2:*276*
external anatomy
2:277, 278–280,
278–280
families 2:277
filmy 2:*276*
internal anatomy
2:277, 281–282,
281–282
lady 2:*276*, 277, *278*
leaves (fronds)
2:278–279
growth 2:282
leptosporangiate 2:*276*
reproductive system
2:277, 283–285,
283–285
rhizomes 2:279–280,
282, 285
stag's-horn 2:*276*, 280
tree 2:279–280
water ferns 2:280
Ferocactus 8:1052
ferret 10:*1324*, 1328
black-footed 10:1328
European 10:*1342*
skull 10:*1333*
fertilization 7:990–991
double 7:1002, *1003*
external 3:445; 7:*992*,
993
in flowering plants
7:1002, *1003*;
10:1313
internal 7:993
fibrin 1:119
fibula 9:*1174*
deer 7:970
fiddleheads 2:282
"fight or flight" response
4:570;
10:1372–1373
fighting for mates
deer 7:982
giant tortoises
9:*1271*

fighting for mates
    (continued)
    giraffes **3**:364
    hares **4**:493
    sea horses **8**:1108
    weasels **10**:1341
fighting for territory,
    wildebeests
    10:*1379*
file snake **3**:*401*
    Arafura **3**:*399*
Filicales **2**:*276*
filter feeders **2**:175
fimbriae (pili) **1**:*47*, 49
fins
    adipose **9**:1284
    anal **8**:1063, 1102
    caudal **8**:1066; **9**:1244,
        1284
    dorsal **2**:190; **8**:*1062,
        1063*–1064, *1073*;
        9:1244
    pectoral **8**:1065, 1066,
        1102; **9**:1174, 1244,
        1249, 1284
    pelvic **8**:1062; **9**:1174,
        1241, 1244
fir **8**:*1138*
firefly **10**:*1344*, 1348,
    1349, 1350–1351
fires, forest, and sequioas
    8:*1149*
firethorn **1**:*36*
Firmicutes **1**:47
fish
    bony **1**:*122*; **3**:*436*;
        8:*1060*, 1061,
        1096, 1097;
        9:*1238, 1280*
    brain **6**:*795*
    cartilaginous **3**:*436*;
        4:*460*, 461; **8**:*1060,
        1096*; **9**:*1238, 1280*
    elopomorph **3**:*436*
    evolution of tetrapods
        from **9**:1158–1159

the fastest **8**:1064
fins **9**:1174
flatfish **9**:1244
gas exchange by
    8:*1012*
hormones
    calcium-regulating
        2:268
    growth-inhibiting
        2:267
intestines **2**:181
jawed **1**:*122*–123;
    4:*446*; **8**:1097
jawless (agnathans)
    3:*436*; **4**:*446*–447,
    460; **8**:*1060, 1096*;
    9:*1238, 1280*
lobe-finned **1**:*122,
    123*; **3**:*436*; **8**:*1060,
    1096*; **9**:*1158*,
    1174, *1280*
with lungs **8**:*1020*
ray-finned **3**:*436*;
    8:*1060*, 1061,
    1096, 1097; **9**:1174,
    1280, 1281
scales **4**:*463*; **9**:*1283*
swim muscles
    6:764–765
teeth **2**:174
fisher **10**:1330, 1342
fission
    binary *see* binary
        fission
    longitudinal **8**:1094
    multiple **1**:24–25
Fissipedia **3**:*410*;
    10:*1382*
flagella **1**:*75*
    of bacteria **1**:49
    of diatoms' sex cells
        2:170, 171
flagellates **2**:*164*, 165
flagellin **5**:610–611
flamingo **5**:695
flask fungi **6**:771

flatworm
    diffusion of oxygen
        8:*1029*
    excretion **2**:175, *187*
    gut **2**:183
    nervous system **6**:*781*,
        792
flea **4**:*538*
flehmen response **4**:519;
    5:676; **7**:976;
    10:1373, *1430*
flesh fly *see* fly, flesh
flexor carpi ulnaris **1**:*15*
flicker
    Andean **10**:1408
    gilded **8**:*1057*
    northern **10**:*1412*
flight
    bat **3**:295, *296*, 297,
        301, 302
    bird **9**:1173
        albatrosses **1**:*12*–13
        buzzards **2**:223
        eagles **2**:222, 231
        hummingbirds
            5:583–584, 586
        vultures **2**:223, 231
        woodpeckers
            10:*1407*
    dragonfly **2**:212, 214
    hawkmoth **4**:*501*
    honeybee **4**:*533*
Florida, waters off,
    jellyfish **5**:645
Florideophycidae **8**:1127
flounder **9**:1241
flowers
    apple tree **1**:*39, 43*;
        7:*1001*
    cactus **8**:*1051,
        1058–1059*
    grass **6**:751–*752*;
        7:*1001*
    maize **6**:*753*
    orchid **6**:*825, 826, 828,
        832*–833

perfect and imperfect
    7:1002
potato **7**:*921, 926, 927*
Venus flytrap
    10:1312–*1313*
fluke, blood *see* blood
    fluke
fly **4**:*538*, 539; **5**:*692*
    aschizan **4**:*538*
    brachyceran **4**:*538*
    cyclorrhaphan **4**:*538*,
        540, 546
    dung **4**:548
    flesh **4**:*538*, 539
    fruit (vinegar) **4**:539,
        543
    head **4**:*546*
        mutations **6**:793
    hover **4**:530, *538*, 539,
        547; **6**:762
    humpback **4**:*538*
    nematoceran **4**:*538*
    parasitic **4**:*538*
    schizophoran **4**:*538*
    stable **4**:539
    tsetse **4**:546, 548, 549
    warble **4**:539, 547
fly agaric **6**:*771*, 772,
    775
flying fox **3**:*293*
    aerodynamics **3**:297
    tail **3**:*296*
flying lemur **3**:*297*
follicle-stimulating
    hormone (FSH)
    2:271; **7**:997–998,
    999
fontanelle **9**:1171
foramen magnum
    4:556
Foraminifera **1**:21, 22;
    2:*164*
foraminiferans **1**:22;
    2:*164*, 165
    cell coverings **2**:167
Forcipulatida **9**:*1224*

forests, pygmy
    hippopotamuses
    4:510
formic acid **1**:28, *29*, 30
Formicidae **1**:*26*
Formicinae **1**:27
fossils **9**:1158
    anteater **3**:315
    eurypterid **8**:1077
    formation **9**:1157
    hagfish **4**:449
    hummingbird **5**:585
    otter **6**:847
    rhinoceros **8**:1036
    seaweed **8**:1130
fovea **2**:229; **9**:1218
fox **6**:*844*; **10**:*1382*, 1387
    arctic **10**:1391
    red **10**:1387
frog **1**:*54*; **6**:*800*
    "ancient" **1**:*54*
    calls **1**:*55*, 57, 61
    common **1**:*54*
    Darwin's **1**:*54*, 63
    edible **1**:*54*
    European tree **1**:*62*
    gold **1**:*54*
    goliath **1**:55
    jumping and
        swimming **1**:*59*
    metamorphosis **1**:*57*;
        2:267
    "new" **1**:*54*
    poison arrow **2**:*265*
    reproduction, external
        fertilization **7**:*992*
    Rio Grande leopard
        1:*54*
    skeleton **9**:1159–1160
    tropical **1**:*54*
    true **1**:*54*, 55
    see also bullfrog
fruit bat **3**:**292–311**
    circulatory and
        respiratory systems
        3:293, *306–307*
digestive and excretory
    systems **3**:293,
    308–309
Egyptian **3**:*305*, 306,
    309
epauletted **3**:*292*, 311
external anatomy
    3:293, *294–297*
gray-headed **3**:293,
    *294, 295, 298, 306,
    308, 311*
flight **3**:297, *301*
nerves inside head
    of **3**:*303*
hammer-headed **3**:*292*,
    311
Indian short-nosed
    3:*309*
leaf-nosed **3**:*294*
long-tongued **3**:*292*
muscular system **3**:293,
    301–302
nervous system **3**:293,
    303–305
New World (microbats)
    3:*292–293, 294*
tail **3**:*296*
Old World (megabats)
    3:*292*, 293, *294*
fruit-eating **3**:*292*,
    293
nectar-eating
    3:*292*, 293
pygmy **3**:293
reproductive system
    3:293, *310–311*
rousette **3**:*292*, 304
skeletal system **3**:293,
    298–300
tube-nosed **3**:*292*
fruit fly *see* fly, fruit
frustule *see* theca
fucoxanin **8**:1127
Fucus **8**:1131, 1135
    vesiculosus **8**:*1126*
fungi **6**:*770*, 770
catching nematodes
    6:*774*
hormones **2**:269
furcula (wishbone) **1**:14;
    7:873, 874

G

Galápagos islands, giant
    tortoises **9**:1270
Galaxiidae **9**:1281
Galictis **10**:*1324*
gallbladder **2**:184; **4**:568
    missing in giraffes
    3:371
gametangia **8**:1137
gametes (sex cells) **1**:67,
    87; **7**:985
    meiosis and **1**:85;
    7:*991*
gametophyte **7**:993,
    1000, 1002
    fern **2**:277, 283, 284,
    285; **7**:*993*
    seaweed **8**:*1134*
ganglia **6**:780–781, 792,
    792–793
    of bivalves **3**:*353*
    of earthworms
    6:792–793
    of giant centipedes
    3:*337*
    of giant clams **6**:*792*
    of hawkmoths **4**:*502*
    of houseflies **4**:*545*
    of humans **4**:563–564
    of scorpions **8**:*1079*
    of tarantulas **9**:1257,
    1259
    of weevils **10**:1354
Ganoderma **6**:*773*
gap junctions **6**:766
gar, scales **9**:1283
gas exchange **8**:*1012*,
    1013–1015
gaster **1**:28, *29*
Gasteromycetidae **6**:*770*
gastrin **2**:272
gastrocnemius muscle
    4:561; **6**:839;
    10:1412
gastrodermis **8**:1091
gastroliths
    in birds **2**:178; **7**:879
    in crocodiles **2**:161
Gastropoda **3**:*342*; **6**:812
gastropods, circulatory
    system **1**:107
gastrulation **7**:1004
gas vesicles, bacterial
    1:51
Gavialidae **2**:*148*
gazelle **10**:*1360*
    forelimbs **3**:*421*
Gecarcinidae **1**:*132*
gecko, skin color **5**:*631*;
    6:*789*
gelada **6**:725
Gelidium **8**:1127
Gempylidae **8**:*1060*
General Sherman tree
    8:1139
genes **7**:984
    alleles **1**:76–79
    and altruism **10**:1401
    Hox **7**:1005
    recessive and dominant
    1:78
    S-genes **1**:45
    "wild-type" **1**:78
genetic disorders,
    immune disorders
    5:*622*, 623
geniohyoid muscles
    10:*1412*
Genlisea **10**:1309
genotypes **1**:77
Gentianales **7**:*916*
geoduck **3**:*342*, 345,
    347
Geomyidae **7**:*946*
geophagia **10**:1434
Geophilomorpha **3**:*332*

Gerbillinae **7**:*946*
gestation **7**:1005
  variable period **3**:330
gharial **2**:149
  false **2**:150
  snout **2**:150, *151*, 153
  teeth **2**:*154*
GHRH **2**:270
giant anteater
    3:**312–331**
  circulatory and
    respiratory systems
    3:313, *329*
  digestive and excretory
    systems **3**:313,
    325–326
  external anatomy
    3:313, *314–317*
  muscular system **3**:313,
    322–324; **6**:*763*
  nervous system **3**:313,
    327–328
  reproductive system
    3:313, *330–331*
  skeletal system **3**:313,
    318–321
giant centipede
    3:**332–341**
  anatomy
    external **3**:333,
    334–335
    internal **3**:333, *336*
  circulatory and
    respiratory systems
    3:333, *338*
  digestive and excretory
    systems **3**:333,
    339–340
giant desert centipede
    3:*333*
giant redheaded
    centipede **3**:*334,
    335, 336, 337, 339,
    341*
  nervous system **3**:333,
    337

reproductive system
    3:333, *341*
giant clam **3**:**342–355**
  anatomy
    external **3**:343,
    344–345
    internal **3**:343,
    346–347
  circulatory and
    respiratory systems
    3:343, *354*; **8**:*1016*
  digestive and excretory
    systems **3**:343, *352*
  muscular system **3**:343,
    350–351
  nervous system **3**:343,
    353
  reproductive system
    3:343, *355*
Giardia lamblia **1**:20
gibberellins **2**:266
gibbon **1**:*88*, 89, 95, 96;
    4:*550*
Gigantornis **1**:14
gills **8**:1016–1017,
    1029–1030
  of crabs **1**:138
  dermal **9**:1227
  of fish **2**:176
    coelacanths **1**:*129*
    gulper eels **3**:*443*
    hagfish **4**:*455*
    hammerhead sharks
    4:*472–473*
    sea horses **8**:*1105*
    sharks **8**:*1027*
    trout **9**:*1289*
    tuna and billfish
    8:*1071*
  of giant clams **3**:*345,
    347, 352, 354*;
    8:*1016*
  of insects **8**:1017
  of octopuses **6**:*821*
  ventilation
    8:1026–*1027*

Gilpichthys **4**:449
ginkgo **8**:*1138*
Ginkgophyta
    8:*1138*
Giraffa
  camelopardalis **3**:*356*
  jumae **3**:361
giraffe **3**:**356–373**;
    7:*964*; **10**:*1360*
  circulatory and
    respiratory systems
    3:357, *368–369*
  digestive and excretory
    systems **3**:357,
    370–371; **10**:*1377*
  external anatomy
    3:357, *358–361*
    Masai **3**:*359*
  muscular system **3**:357,
    364–365
  nervous system **3**:357,
    366–367
    Nubian **3**:*359*
  reproductive system
    3:357, *372–373*
  reticulated **3**:*358*, 359
  skeletal system **3**:357,
    362–363; **9**:*1173*
  teeth **2**:*174*
Giraffidae **3**:*356*
giraffids **3**:*356*, 357, 361
Giraffoidea **3**:*356*; **7**:*964*;
    10:*1360*
gizzard
  of birds **2**:178
    albatrosses **1**:18
    eagles **2**:230
    woodpeckers
    10:1415
  of crocodiles **2**:161
  of invertebrates **2**:179
glenoid fossa **10**:1411
glial cells **6**:785
glires **4**:*476*, 477
glochids **8**:1053
glomeruli **8**:1045

glossa **4**:530
glottis **8**:1019
glowworm **10**:1348,
    1349, 1350
glucagon **2**:*272*, 273;
    4:*571*
glucocorticoids **2**:272
glumes **6**:751
gluteal muscles **4**:561
gluteus maximus **4**:561
gluteus medius **4**:561
glycogen **5**:680;
    8:1106–1107
glycoproteins **10**:1318
glyptodonts **3**:320
Gnathostomata **1**:*122*;
    3:*436*; **4**:*446, 460*;
    8:*1060, 1096*;
    9:*1238, 1280*
gnathostomes
    1:*122*–123; **3**:*436*;
    4:*460*; **8**:*1060,
    1096*, 1097;
    9:*1238, 1280*
Gnetophyta **8**:*1138*
gnetophyte **8**:*1138*
goat **3**:*356*, 357; **7**:*964*;
    10:*1360*
goblet cells **2**:*181*; **4**:568
Golgi apparatus (Golgi
    body) **1**:*71*
  of diatoms **2**:169
gonadotropin-releasing
    hormone (GnRH)
    7:*997*–998; **10**:*1416*
gonadotropins **2**:269, 271
Gongora **6**:833
gonopore **1**:140; **3**:341
  of earthworms **2**:237,
    243
gopher, pocket **7**:*946*
gorilla **1**:*88*, 89, 95, 103;
    4:*550*
  teeth **4**:569
Gorilla **4**:*550*
  gorilla **1**:*88*

Gossypium **6**:752
gracilis muscles **8**:1111
grafting, apple trees
   1:*45*
Gramineae **6**:*744–745*
grana **6**:*750*
granulosa cells **7**:995
grapevine **1**:*36*
grass **6**:*744–745*
   marsh *see* marsh grass
   pampas **6**:751
grasshopper, nervous
   system **6**:*781*
grayback **2**:*208*
grayling **9**:*1280*
gray whale **3**:*374–395*
   circulatory and
     respiratory systems
     3:375, *389–391*
   digestive and excretory
     systems **3**:375,
     392–393
   external anatomy
     3:375, *376–379*
   eyes **3**:387
   muscular system **3**:375,
     383–384
   nervous system **3**:375,
     385–388
   reproductive system
     3:375, *394–395*
   skeletal system **3**:375,
     380–382
Great Barrier Reef, sea
   anemones **8**:*1091*
green anaconda
   3:*396–409*
   circulatory and
     respiratory systems
     3:397, *407*
   digestive and excretory
     systems **3**:397, *408*
   external anatomy
     3:397, *398–401*
   muscular system **3**:397,
     405

   nervous system **3**:397,
     406
   reproductive system
     3:397, *409*
   skeletal system **3**:397,
     402–404; **9**:*1175*
grison **10**:*1324, 1327*,
   1329
grizzly bear **3**:**410–429**
   circulatory and
     respiratory systems
     3:411, *424–425*
   digestive and excretory
     systems **3**:411,
     426–427
   external anatomy
     3:411, *412–415*
   muscular system **3**:411,
     420–421
   nervous system **3**:411,
     422–423
   reproductive system
     3:411, *428–429*
   skeletal system **3**:411,
     416–419
growth, allometric **3**:331
guanin **9**:1284
guard cells **1**:42
Guatemala, Montezuma
   pine **8**:*1140*
guenon **6**:*724*, 725
   hands and feet **6**:*728*
guitarfish **9**:*1238*, 1243
gull, wing bones **7**:*874*
Gulo **10**:*1324*
gulper eel **3**:**436–445**
   circulatory and
     respiratory systems
     3:437, *443*
   common Atlantic
     3:*437, 438, 441,*
     444, 445
   digestive and excretory
     systems **3**:437, *444*
   external anatomy
     3:437, *438–440*

   nervous system **3**:437
   reproductive system
     3:437, *445*
   skeletal and muscular
     systems **3**:437, *441*
Gymnocladium hossei
   8:*1050*
Gymnophiona **1**:*54*;
   6:*800*
Gymnospermae **1**:*36*;
   6:*824*; **7**:*916*;
   8:*1048, 1138*
gymnosperms **1**:*36*, 37;
   6:*744, 824*; **7**:*916*,
   1000; **8**:*1048,*
   1138; **10**:*1300*
   reproduction in **7**:1003
gyri **4**:565; **6**:*737*

H

hagfish **4**:**446–459**
   Atlantic **4**:*446, 449,*
     452, 453, 456, 459
   circulatory and
     respiratory systems
     1:111, *112*; **4**:447,
     455–456
   digestive and excretory
     systems **4**:447,
     457–458
   external anatomy
     4:447, *449–451*
   nervous system **4**:447,
     453–454
   reproductive system
     4:447, *459*
   skeletal and muscular
     systems **4**:447, *452*;
     9:*1159*
hair
   bear **3**:*412–413*
   defensive hairs,
     tarantula **9**:*1256,*
     1260
   guard hairs **6**:849;
     10:*1385*

   hare **4**:*481*
   human **4**:*554*
   otter **6**:*847–848*, 849
   pubic **4**:554
   puma **7**:933
   secondary hairs (awns)
     10:1386
Haliaeetus leucocephalus
   2:*220*, 221
Halictidae **4**:*526*
hallux, opposable **6**:729
halteres **2**:211; **4**:*540*
hammerhead shark
   4:**460–475**
   circulatory and
     respiratory systems
     4:461, *472–473*
   digestive and excretory
     systems **4**:461, *474*
   external anatomy
     4:461, *462–463*
   gills **4**:*472–473*
   great **4**:*461*
   muscular system **4**:461,
     466–467
   nervous system **4**:461,
     468–471
     electroreceptors
       6:*787*
   reproductive system
     4:461, *475*; **7**:997
   scalloped **4**:*462*
   skeletal system **4**:461,
     464–465
hamster **7**:946
   tail **7**:955
hamstring muscles **4**:561
hamuli **1**:29; **4**:529
haploid cells **1**:76, 87;
   7:985, 990
haptera **8**:*1129, 1131*
hare **4**:**476–493**
   arctic **4**:*479, 481, 489*
   brown **4**:477,
     478–479, 480, 482,
     483, 484, 489, 490

hare *(continued)*
  gestation period 4:*492*
    young 4:*493*
  circulatory and
    respiratory systems
    4:*477, 487–488*
  digestive and excretory
    systems 4:*477,
    489–491*
  external anatomy
    4:*477, 478–481*
  mountain 4:*481*
  muscular system 4:*477,
    484–485*
  nervous system 4:*477,
    486*
  reproductive system
    4:*477, 492–493*
  skeletal system 4:*477,
    482–483*
  snowshoe 4:*479, 481*
hartebeest, hooves
    10:*1366*, 1367
harvestman 8:*1074*, 1075
hatchetfish 3:440
haustellum 5:697
haversian canals 9:1168
Hawaiian islands,
    chameleons 5:624
hawk 2:*220*
hawkmoth 4:**494–505**
  anatomy
    external 4:495,
    496–499
    internal 4:495,
    500–501
  circulatory and
    respiratory systems
    4:495, *503*
  death's-head 4:*499*
  hummingbird 4:*498*
  nervous system 4:495,
    502
  poplar 4:*495*
  reproductive system
    4:495, *504–505*

hawthorn 1:*36*, 37, 38
hay fever 5:*613*
hearing 6:*789*
heart
  of amphibians 1:*110,
    111–112*
    bullfrogs 1:*61*
  of arthropods
    honeybees 1:*117*
    spiders 9:1261
  of birds 1:*110*, 115
  of fish 1:*110–111*;
    9:1248
    coelacanths 1:129
    hagfish 1:111, *112*;
    4:*456*
    sharks and rays 4:473;
    9:1248
heartbeat 1:*114,
    115–117*; 6:*766*,
    767
  of mammals 1:112;
    4:487
    bats 3:301
    elephants 2:258, *259*
    humans 1:*113–117*;
    4:*567*
  neurogenic 6:767
  of reptiles 1:*110*, 112
    chameleons 5:*632*
    crocodiles 1:*112*;
    2:159
    snakes 3:407
    tortoises 9:*1275*,
    1276
    turtles 9:*1203*
  transplants 5:*617*
  valves 1:*113*, 116
  "venous hearts" 1:*115*;
    3:307
  of vertebrates
    1:*110–112*
    cardiac muscle 1:67,
    115–116; 2:254;
    6:755, *766–767*;
    7:954; 10:1338

heart attacks 1:115;
    4:567
heat exchangers,
    countercurrent *see*
    countercurrent heat
    exchangers
hectocotylus 6:817, *823*
Heliamphora 10:1301
Helianthus annus 6:752
Heliozoa 2:*164*
heliozoans 1:21; 2:*164*,
    167
hellbender 6:801, *809*
hemal arches 9:1172
Hemiascomycetes 6:*770*
Hemigaleidae 4:*460*
hemipenis 5:*634*;
    7:*997*
hemocoel 3:332; 4:500;
    5:684, 696
  of ants 1:30
  of centipedes 3:336
  of scorpions 8:1080
  of weevils 10:*1355*
hemocyanin 1:138;
    3:338; 5:688; 6:821;
    8:1015, 1027;
    9:1261
hemoglobin 1:108, *109*,
    119; 4:456; 5:678;
    8:1015, *1023–1025*
    albatross 1:17
    earthworm 2:241
    elephant 2:259
    fetal 8:1024
    otter 6:857
    penguin 7:878
    platypus 7:893
hemolymph 3:332; 5:684
  ant 1:30
  centipede 3:334, 336,
    338
  dragonfly 2:213, 216,
    217
  hawkmoth 4:500, *503*
  honeybees 4:535

housefly 4:*544, 547,*
    548
  ladybug 10:*1348*
  louse 5:696
  scorpion 8:1080
  spider 9:1261
  weevil 10:*1352*, 1353,
    1355
hepatic portal system
    1:116
hepatic portal vein 5:588
hepatic sinus 8:1120
hepatopancreas
    8:*1106–1107*
herbivores
  intestines 2:181
  teeth 2:174
hermaphroditism 7:*998*
  sea anemones and
    8:1095
Herpestidae 5:*592, 664*
Herpesviridae 10:1314
herpesvirus 10:*1322*
Hesperoidea 4:*494*, 495
Heterobasidiomycetes
    6:*770*, 771
heterocysts 1:*49*
Heterodonta 3:*342*, 343
heterotrophs 1:66
  bacteria as 1:52
Hexapoda 1:*26*; 2:*208*;
    3:*332*; 4:*494*;
    5:*692*; 10:*1344*
hexapods 1:*26*; 2:*208*;
    3:*332*; 4:*494–495*,
    526, 538; 5:*692*;
    10:*1344*
Hexaprotodon liberiensis
    4:*506*, 510
hibernation
  bears
    grizzly 3:425, 427,
    428, 429
    polar 3:*425*
  tortoises 9:1276
Hippocampinae 8:*1096*

hippocampus **4**:565;
  6:798
Hippopotamidae **4**:*506*
hippopotamus **3**:*356*,
  357; **4**:**506–525**;
  7:*964*; **10**:*1360*
  circulatory and
    respiratory systems
    4:507, *520–521*
  common **4**:*506*, 508,
    *509, 510, 512, 514,
    516*, 517, *518*, 519,
    520, 522, 523,
    524–525
  digestive and excretory
    systems **4**:507,
    522–523
  external anatomy
    4:507, *508–511*
  muscular system **4**:507,
    516–517
  nervous system **4**:507,
    518–519
  pygmy **4**:*506*, *509*,
    510, 523
  reproductive system
    4:507, *524–525*
  skeletal system **4**:507,
    512–515
Hippopotamus amphibius
    4:*506*
Hippotraginae **10**:*1360*,
    1361
Hirudea **2**:*234*
histamine **5**:*613*
HIV (human
    immunodeficiency
    virus) **5**:*616*, 620;
    10:*1315*, *1323*
hog **4**:*506*
holdfast **8**:1128, *1129*
Holothuroidea **9**:*1224*,
    1225
Homarus americanus **5**:*684*
homeostasis, hormones
    and **2**:268

Hominidae **1**:*88*, 89;
    4:*550*
hominids **4**:*550*, 551
  extinct, diet **4**:557
  foramen magnum
    4:556
hominines **4**:*550*, 551
Hominoidea **1**:*88*; **4**:*550*
Homo **4**:*550*, 551
  erectus **4**:*550*
  floresiensis **4**:551
  heidelbergensis **4**:*550*
  neanderthalensis **4**:*550*
  sapiens **1**:*88*; **4**:*550*
Homobasidiomycetes
    6:*770*, 771
homodontia **1**:62
Homoptera **5**:*692*
honeybee **1**:*26*;
    4:**526–537**
  anatomy
    external **4**:527,
    528–531
    internal **4**:527,
    532–533
  Asian dwarf **4**:*526*
  Borneo **4**:*526*
  circulatory and
    respiratory systems
    1:*117*; **4**:527, *535*
  digestive and excretory
    systems **4**:527, *536*
  dwarf **4**:*526*
  eastern **4**:*526*
  giant **4**:*526*
  Himalayan **4**:*526*
  nervous system **4**:527,
    534
  reproductive system
    4:527, *537*
honeycomb **4**:529
honeydew **1**:33
hooves
  giraffe **3**:*363*
  hippopotamus **4**:*510*,
    513–514

zebra **10**:*1421*–1422
hormones **2**:264
  affecting growth and
    development **2**:265,
    266–267
  fungal **2**:269
  gastrointestinal (GI)
    2:272
  half-life **2**:271
  and homeostasis **2**:265,
    268
  lipid-insoluble **2**:*264*,
    265
  lipid-soluble **2**:265,
    271
  mode of action **2**:*264*,
    265
  plant **2**:266
  and reproduction
    2:265, 269
  sex, of vertebrates
    7:996–999
  types **2**:265
horns **7**:*972*; **9**:1171;
    10:1364–*1365*,
    1367, 1369
  females with **10**:1379
hornworm, tobacco
    4:497
horse **8**:*1032*; **10**:*1418*,
    1419
  Clever Hans **10**:1431
  domestication **10**:1423
  geophagia **10**:1434
horse-ass hybrids
    10:1437
hybrids **10**:1437
  jaw **10**:*1425*
  Przewalski's **10**:*1418*,
    1423
  sounds **10**:1433
  teeth **10**:1435
  ventilation at speed
    8:1031
horsetail **2**:*276*, 279, 284
  hairy **2**:279

houndshark **4**:*460*
housefly **4**:**538–549**
  anatomy
    external **4**:539,
    540–543
    internal **4**:539, *544*
  circulatory and
    respiratory systems
    4:539, *547*
  digestive and excretory
    systems **4**:539, *548*
  feet **4**:*542*
  halteres **2**:211; **4**:*540*,
    542
  maggots (larvae) **4**:547,
    548, *549*
  nervous system **4**:539,
    545–546
  reproductive system
    4:539, *549*
  sucking liquids **4**:*541*,
    548
hover fly *see* fly, hover
human **1**:*88*, 89;
    4:**550–573**
  bipedalism **4**:552–553,
    556, 558, 561
  circulatory and
    respiratory systems
    4:551, *566–567*
    ventilation **8**:*1028*
  digestive and excretory
    systems **4**:551,
    568–569
    liver **2**:*184*; **4**:568
    stomach **2**:177
  embryo **7**:*1004*
  endocrine system
    2:270; **4**:*570–571*
  external anatomy
    4:551, *552–554*
  heart **1**:*113*–117;
    4:*567*
  heart attacks **4**:567
  life at high altitudes
    4:566

human *(continued)*
  muscular system **4**:551,
    559–561
  nervous system **4**:551,
    562–565
    brain **6**:*794, 795,
      796, 799*
    brain size and
      intelligence **4**:*564*
    eyes **4**:*553*; **6**:*788*
    vision **4**:*563*
  oral cavity **2**:*173*
  reproductive system
    4:551, *572–573*;
    7:*994*–996
  skeletal system **4**:551,
    555–558; **9**:*1169*
    hand bones **9**:*1174*
  tears **2**:*275*
  teeth **4**:*557*
  urinary system, male
    2:*185*
  vision **4**:*563*
  vocal cords and
    muscles **4**:*560*
  whites of the eyes
    4:*553*
Human Genome Project
  1:*77*
human
  immunodeficiency
  virus *see* HIV
humerus
  albatross **1**:*14*
  eagle **2**:226
  human **4**:558
  seal **8**:1116
  whale **2**:196
hummingbird
  5:**580–591**;
  8:1059
  Anna's **5**:*584*, 590
  black-chinned **5**:*583*
  circulatory and
    respiratory systems
    5:581, *588*

digestive and excretory
  systems **5**:581,
  589–590
external anatomy
  5:581, *582–584*
flight **5**:*583–584*, 586
giant **5**:582
hermit **5**:*580*, 581
muscular system **5**:581,
  585
nervous system **5**:581,
  587
reproductive system
  5:581, *591*
ruby-throated **5**:*580*,
  581, *582*, 584, *585*,
  586, 588, 590, 591
skeletal system **5**:581,
  585
sword-billed **5**:589
Hyaena hyaena **5**:*592*
Hyaenidae **5**:*592, 664*;
  7:*928*
Hyaeninae **5**:*592*
Hyatellidae **3**:*342*
hybrids, equine **10**:1437
hydathodes **6**:748
hydra **1**:106; **2**:183
  nervous system **6**:*781*,
    792
  reproduction, asexual
    7:*987–988*
Hydrobatidae **1**:*10*
Hydrochaeridae **7**:*898*
Hydrozoa **5**:*636*; **8**:*1084*
hydrozoans **5**:*636*, 637,
  638, 639, *640*;
  7:986, 987;
  8:*1084*
hyena **5**:**592–605**, *664*;
  7:*928*; **10**:*1324*
  African folklore and
    5:601
  ancestors **5**:*595*
  brown **5**:*592*, 593, *595*,
    602, 604

circulatory and
  respiratory systems
  5:593, *602*
digestive and excretory
  systems **5**:593, *603*
external anatomy
  5:593, *594–595*
hunting by **5**:600
laughing **5**:602
muscular system **5**:593,
  598–599
nervous system **5**:593,
  600–601
reproductive system
  5:593, *604–605*
scent **5**:595, *600–601*
skeletal system **5**:593,
  596–597
  hind limb bones
    9:*1175*
spotted **5**:*592*, 593,
  595, 596, 597, 598,
  601, 602, 603, *604*,
  605
striped **5**:*592*, 593,
  594–595, 596, 597,
  598, 600, 602, 603,
  604
see also aardwolf
Hylobatidae **1**:*88*, 89;
  4:*550*
Hylocereus undutus **8**:1050
hymen **7**:996
Hymenomycetidae
  6:770
Hymenophyllaceae
  2:*276*
Hymenophyllum **2**:*276*
Hymenoptera **1**:*26*, 27;
  4:*494, 526*, 527
hymenopterans **4**:*526*,
  527
  narrow-waisted **4**:*526*
Hynobiidae **6**:801
hyoglossus muscle **5**:633
hyoglossus tendon **5**:*630*

hyoid apparatus
  of birds **10**:*1411*
    hummingbirds
    5:589
  of elephants **2**:253
  of small cats **7**:*935*
  and the tongue **3**:319,
    380
hyoid arch **9**:1171
hypanthium **1**:*43*
hypersensitivity,
  immediate **5**:616
hyphae **6**:772, 776
hypocalcin **2**:268
hypocingulum **2**:*167*
hypodermis **8**:*1055*
hypopharyngeal gland
  4:532
hypopharynx **4**:540–541
hypophysis **2**:266
hypothalamus **2**:270;
  4:565; **5**:*634*; **6**:798
  effect on the pituitary
    2:266, 270; **4**:*571*;
    6:791; **7**:*997–998*
  of manatees **5**:709
hypovalve, of diatoms
  2:166
Hypsignathus spp. **3**:*292*
Hyracoidea **2**:244
Hyracotherium **10**:*1421*,
  1424, 1426
hyrax **2**:244, 245, 249
Hystricidae **7**:*898*
Hystricognathi **7**:*898*,
  946

I

icefish **8**:1025
Ichneumonidae **1**:*26*
ichthyosaur **3**:379
Ichthyostega **9**:1159
ictithermium **5**:*595*
iguana
  marine, salt-secreting
    glands **2**:161

scales **5**:*627*
teeth **2**:*174*
ileocecal valve **9**:1277
ileum
  insect **2**:183
  toothed whales **2**:205
  vertebrate **2**:181, 182
iliocostal (iliocostalis)
    muscles **2**:156;
    8:1117
immune, defense, and
    lymphatic systems
    5:*606–623*
immune disorders
    5:*622–623*
immunity **5**:618, 621
  adaptive **5**:606, 608,
    612–617
  innate **5**:606, 608,
    609–611
  lymphatic system
    4:488; **5**:606,
    618–621
  see also autoimmunity
immunoglobulins (Igs)
    5:612, 613
  IgE **5**:615
  surface **5**:615, 620
  see also antibodies
impala **10**:*1367*
implantation, delayed
    6:860; **10**:1342
inbreeding, by grizzly
    bears **3**:429
inclusion granules,
    bacterial **1**:51
incus **4**:555; **9**:1170
India
  chameleons **5**:624
  hyenas **5**:593
Indian Ocean, sea horses
    8:*1098*
Indricothelium **8**:*1036*
infanticide, lions **5**:682
inflammation **5**:606–608,
    611

inflation, chameleons
    5:632
influenza
  pandemics **5**:621
  virus **10**:*1317*
Influenzavirus
    10:1314
infraorbital foramen
    7:952
infrasound
  giraffe **3**:366
  hippopotamus **4**:519
  manatee **5**:710
infundibulum **5**:591
Iniidae **2**:*188*
ink, of squid, cuttlefish,
    and octopuses
    6:*817*
Insecta **1**:*26*; **2**:*208*;
    3:*332*; **4**:*494, 526,
    538*; **5**:*692*; **8**:*1074*;
    10:*1344*
insects **1**:*26*; **2**:208–209;
    4:495, 526, *526*,
    538; **5**:*692*; **8**:*1074*;
    10:*1344*, 1345
  circulatory system
    1:107
  digestive system **2**:179,
    183
  evolution **10**:1348
  excretion by **2**:186
  hormones **2**:267, 269
  immunity **5**:608
  largest **2**:217; **8**:*1021*
  molting **2**:267
  skeletal system
    9:1166–1167
  wing movement **6**:*762*
insulin **1**:70; **2**:272, 273;
    4:*571*; **9**:1291
  from genetic
    engineering **1**:50
intelligence, relative
    4:564
interbreeding, deer **7**:981

intercalated disks **6**:766
interclavicle **7**:888–889
intercostal muscles
    3:329; **8**:1028;
    10:1412
interferon **5**:609–610,
    614
interneurons **2**:256;
    3:366; **6**:*783*; **7**:*908*;
    8:1042; **10**:1336
intertransversales muscle
    2:228
intervertebral disks
    9:1172
intestines **2**:180–183;
    7:*942*
  and diet **3**:325
  of eagles **2**:230
  of toothed whales
    2:204, 205
intracranial joints **1**:*126*
intramandibularis muscle,
    crocodile **2**:156
invertebrates
  digestive systems
    2:178–179
    intestines **2**:183
    mouthparts **2**:175
  excretory systems
    2:187
  nervous system **6**:781,
    792–793
Ireland, weasels **10**:1340
iridocytes **6**:816; **9**:1284
ischial callosities **6**:730
Ischnocera **5**:*692*, 693,
    694, 699
Ischyromys **7**:949
islands, animals on
    4:511
Isoptera **1**:*26*
Istiophoridae **8**:1060
Istiophorus
  albicans **8**:*1060*
  platypterus **8**:*1060*

J
jackal **10**:*1382*, 1384
  black-backed
    10:*1382*
jackrabbit **4**:*476*
  antelope **4**:*479*
  black-tailed **4**:*479*
Jackson's chameleon
    5:*624–635*
Jacobson's organ *see*
    vomeronasal organ
jaguar **5**:*664*
  eyes **5**:*668*
jejunum **2**:181, 182
jellyfish **5**:*636–647*
  anatomy
    external **5**:637,
      639–640
    internal **5**:637,
      641–642
  bell **5**:*638*
  box **5**:636, 638, 640,
    643; **8**:*1084*
  common moon **5**:*636*,
    638, *639, 641*, 642,
    643, *644*, 646–647
  digestive and excretory
    systems **5**:637,
    644–645
  dunce-cap **5**:*643*
  lacking blood **3**:354
  lion's mane **5**:*637*,
    640
  moon **5**:636, 638, *639*,
    641, 643, *644*, 646,
    646–647; **7**:*987*
  nervous system **3**:353;
    5:637, 641, *643*;
    6:*789*
  purple stinger **5**:*645*
  reproductive system
    5:637, *646–647*
  skeletal system **5**:642;
    9:*1162*
  true **5**:*636*; **8**:*1084*
jelly fungi **6**:771

jerboa **7**:*946*, 951
  tail **7**:955
Juncaceae **6**:*744*
juniper, Utah **8**:1144
juvenile hormone (JH)
  2:267; **4**:505
Jynginae **10**:*1404*, 1405

K

kangaroo **5**:**648–663**
  ancestors **5**:652
  circulatory and
    respiratory systems
    5:649, *658*
  digestive and excretory
    systems **5**:649,
    659–660; **10**:*1377*
  eastern gray **5**:*648*,
    649
  external anatomy
    5:649, *650–653*
  gray **5**:*651*, *657*
  movement **5**:*652*, 653,
    656
  muscular system **1**:*56*;
    5:649
  nervous system **5**:649,
    657
  red **5**:*648*, 649, *650*,
    651, *652*, *654*, *655*,
    656, 660, 662–663
  reproductive system
    5:649, *661–663*;
    7:*996*
  skeletal system **5**:649,
    654–655
    hind limb bones
      9:*1175*
  tree **5**:649, *660*
  western gray **5**:649
karyogamy **6**:777–778
katoikogenic
  development
  8:1083
kauri **8**:1142
keel **1**:*14*; **7**:873

kelp **8**:*1129*, *1131*
  bullwhip **8**:1130, 1131
  feather boa **8**:*1129*,
    1130
  giant **8**:*1127*, *1132*
  sea urchins and **6**:859
  winged **8**:1130
Kenya, chameleons
  5:624
keratin **10**:1366
kidneys **2**:185, 186;
  8:1045
  lobular/reticulate
    8:1123
kinases, cyclin-dependent
  1:85
kinetoplast **1**:21
Kinixys **9**:1268, 1273
kite **2**:*220*
kiwi **6**:*834*, 835, 840
klipspringer, hooves
  10:*1366*
koala **5**:*648*
Kogiidae **2**:*188*
krait, venom **6**:783
krill **1**:16, 133
kudu **10**:1361
  horns **10**:*1365*

L

labellum (of flowers),
  orchids **6**:832, *833*
labellum (of insects)
  houseflies **4**:*541*
  mosquitoes **4**:*546*
labial palps **3**:*352*
labia major **7**:996
labia minor **7**:996
labium **4**:540–541; **5**:697
labrum **4**:540–541; **5**:697
labyrinthodonts **1**:126
Lacertilia **3**:*396*
lacewing **4**:*538*
lachrymal glands **2**:275
lactate dehydrogenase
  7:875, 878

lactic acid **5**:678; **7**:875,
  878; **8**:1012, 1121
ladybug **10**:*1344*, *1348*,
  1356
Lagomorpha **4**:*476*
lagomorphs **4**:*476–481*,
  484, 486, 489–491,
  492–493
laguno **8**:1125
Lamiales **7**:*916*
Laminaria **8**:1131, 1133
lamp-mussel **3**:355
lamprey **4**:*446–4478*
  hemoglobin **4**:456
  mouth **4**:*458*
  vertebrae **9**:1172
lancelet **4**:454
  hepatic cecum **2**:184
  hormones **2**:269
Langerhans, islets of
  4:571
langur *see* monkey, langur
lanternfish **3**:439, 440
larch **8**:1140
  bark **8**:*1141*
  cones **8**:*1143*
  leaves **8**:*1142*
larvae
  ammocoete **4**:448
  bipinnaria **9**:*1236*,
    1237
  brachiolaria **9**:*1236*,
    1237
  of crabs **1**:141
  of honeybees **4**:536,
    537
  of houseflies **4**:547,
    548, *549*
  leptocephalus **3**:445
  of lice **5**:699
  of newts and
    salamanders **6**:*804*
  planula **5**:*646–647*;
    7:*986*; **8**:*1094*–1095
  of sea anemones
    8:*1094*–1095

  of starfish **9**:*1236*, 1237
  trochophore **3**:355
  veliger **3**:355
  zoea **1**:141
larynx **2**:176; **10**:1414
Lasiocampidae **4**:*494*
lateral line **1**:125, *128*;
  3:*442*; **4**:469;
  6:*790*; **9**:*1288*
  of newts **6**:*807*
  of octopuses **6**:820
  of rays **9**:1247
  of sailfish **8**:1068, *1069*
lateral masseter muscle
  9:*1216*
Latimeria
  halumnae **1**:*122*
  menadoensis **1**:*122*
Latimeriidae **1**:*122*
latissimus dorsi muscles
  1:*15*; **4**:561; **8**:1041;
  10:1370
laver, purple **8**:*1126*,
  1127, *1129*
leaves
  apple tree **1**:39, *41–42*
  grass **6**:746, 747
  internal structure
    6:*830*
  orchid **6**:827–828,
    829–830
  potato **7**:918, *919*
  Venus flytrap **10**:1302,
    1306–1307
leech **2**:234, 235
  mouthparts **2**:236
lemma **6**:751, *752*, 753
lemming **7**:*946*
lemur **1**:95; **4**:550
  flying **3**:*297*
Lentibulariaceae
  10:*1300*–1301
lenticels **7**:922
Lentivirus **10**:1314
leopard **5**:*664*, 671
  clouded **5**:665; **7**:932

Lepdogalaxiidae **9**:1281
Lepidoptera **4**:*494, 538*
lepidopterans **4**:495, 496, 504
Lepidosirenidae **1**:*122*
Lepidosireniformes **1**:*122*
Leporidae **4**:*476*
leporids **4**:477
Leptochariidae **4**:*460*
Leptochloa fusca **6**:748
Leptodactylidae **1**:*54*
Lepus **4**:*476*
leukocytes **5**:611
levator caudae muscle **2**:228
levator coccygeus muscle **2**:228
Leydig cells **7**:995
lichen **6**:773
Lieberkühn, crypts of **4**:*569*
life, beginnings of **1**:64, 69
ligamentum nuchae **3**:364
light production *see* bioluminescence
lignin **1**:42
Liliaceae **6**:*744*
Liliales **6**:*824*
Liliidae **6**:825
lily **6**:*744*
  cobra **10**:1301, 1305
limbic system **6**:798
limbs, pentadactyl **9**:*1174–1175*
lion **5**:603, **664–683**
  African **5**:665, *670, 671, 672, 674, 676, 677, 679, 681, 682*
  Asiatic **5**:665
  Barbary **5**:667
  circulatory and respiratory systems **5**:665, *677–678*

cubs **5**:682, *683*
digestive and excretory systems **5**:665, *679–680*
evolution **5**:668
external anatomy **5**:665, *666–669*
hunting by **5**:668–669, *672–673, 674–675, 678, 680*
muscular system **5**:665, *672–673*
nervous system **5**:665, *674–676*
prides (groups) of **5**:*680*, 681, 682
reproductive system **5**:665, *681–683*
roaring **5**:*671*
skeletal system **5**:665, *670–671*
lionfish, venom **2**:275
Lipotidae **2**:*188*
lips **2**:173
Lithobiomorpha **3**:*332*
liver **1**:116; **2**:182, 184; **5**:680; **8**:*1045*, 1106–1107
lizard **3**:*396*; **5**:*624*, 625; **9**:*1194*
  alligator, scales **2**:151
  color change **6**:*789*
  limbless **3**:407
  red blood cells **4**:*521*
  sungazer, pheromones **2**:274
  tail that can drop off **5**:629
  vision **9**:*1274*
  whiptail, asexual reproduction **7**:*986*, 989
llama **7**:*964*; **10**:*1360*
lobster **1**:133; **5**:**684–691**
  American **5**:*684*, 685, *686, 687, 689, 691*

external anatomy **1**:135; **5**:685, *686–687*; **9**:*1167*
internal anatomy **5**:685, *688–689*
clawed **5**:685
claws **2**:175; **5**:*687*
  muscles **6**:762
heart **6**:767
nervous system **5**:685, *690*
pincer **5**:*684*
reproductive system **5**:685, *691*
rock **5**:*684*
slipper **5**:*684*
spiny **5**:*684*
true **5**:*684*
lodicules **6**:*752*
longissimus caudalis muscles, crocodile **2**:157
longissimus muscles **7**:936–937; **8**:1117
longitudinal fission **8**:1094
longus colli **10**:1412
Lontra canadensis **6**:*844*
Lophophora **8**:1052
loris **1**:95; **4**:*550*; **6**:738
louse **5**:**692–699**
  anatomy
    external **5**:693, *694–695*
    internal **5**:693, *696*
  digestive and excretory systems **5**:693, *697–698*
  human body **5**:693, *698*, 699
  human head **5**:*693, 694*, 699
  pubic (crab) **5**:693, *695*
  reproductive system **5**:693, *699*
  shaft **5**:693

true lice **5**:*692*
whale (cyamid) **3**:*378*
Loxodonta **2**:244
  africana **2**:*244*
  cyclotis **2**:*244*
luciferase **10**:1349
luciferins **3**:439; **10**:1349
lugworm **8**:1026
Lumbricidae **2**:*234*
lumbricid worms **2**:*234*
Lumbricus terrestris **2**:*234*, 235
lungfish **1**:123; **8**:*1020*
  African **1**:*122*, 129
  Australian **1**:*122*, 129; **8**:*1020*
  South American **1**:*122*, 129
lungs **8**:1030
  of birds **8**:1020
    eagles **2**:231
  book **8**:1015, *1022, 1080*; **9**:*1257, 1261*
  of elephants **2**:259
  evolution **1**:129
  of snakes **3**:407
luteinizing hormone (LH) **2**:269, 271; **4**:571; **7**:998, 999
Lutra lutra **6**:*844*
Lutrinae **10**:*1324*
Lybia tessellata **1**:135
Lycoperdon pyriforme **6**:*778*
Lycophyta **2**:276
Lycosa tarantula **9**:*1253*
lymph **4**:488; **5**:608, 618, *619*
lymphatic system *see* immune, defense, and lymphatic systems
lymph nodes ("glands") **5**:618, *619*
lymphocytes **1**:*108*, 109; **4**:488; **5**:608, 621

lymphocytes *(continued)*
  B (B-cells) **5**:612, *615*, 617, *618–619*, 620–621
  CD4 **5**:*616*; **10**:*1315*, 1323
  helper T-cells **5**:616, 617, 619
  production of **5**:*618*–619
  T (T-cells) **5**:*615*, 616, 617, *618*, 619
lynx **7**:*928*
lyriform organs **9**:1260
lysosomes **1**:73
lysozyme **5**:608, 611
Lyssavirus **10**:1314

M

Macaca **6**:*724*
macaque **6**:*724*, 725, *728*
  Barbary **6**:725, 727
  brain **6**:737
Macroglossinae **3**:*292*
macrophages **5**:*610*, 611
Macropodidae **5**:*648*, 649
Macropus **5**:649, 653
  giganteus **5**:*648*
  robustus **5**:*648*
  rufogriseus **5**:*648*
  rufus **5**:*648*
mad cow disease **10**:1318
madreporite **9**:1227, 1229
maggots, rat-tailed **4**:547
magnetic fields, detection **6**:791; **9**:1202
Maihuenia **8**:1050
Maihuenioideae **8**:*1048*, 1049
Maihueniopsis glomerata **8**:1052
maize (corn) **6**:*749*, *753*

Majidae **1**:*132*
major histocompatibility complex (MHC) **5**:617, 618, 622
Malacostraca **1**:*132*, 133; **5**:*684*, 685
malacostracans **1**:*132*, 133; **5**:*684*, 685
malaria **5**:*607*
Malaxis paludosa **6**:832
Malay peninsula, porcupines **7**:900
malleus **4**:555; **9**:1170
Mallophaga **5**:692
Malpighian tubules **2**:187; **4**:536
  of ants **1**:33
  of centipedes **3**:336
  of houseflies **4**:548
  of lice **5**:*697*, 698
  of spiders **9**:1257, 1263
  of weevils **10**:1353
Malus **1**:37
  domestica **1**:*36*
Mammalia **1**:*88*; **2**:188, *244*; **3**:*312, 356, 374, 410*; **4**:*476, 506, 550*; **5**:*592, 648, 664, 700*; **6**:*724, 844*; **7**:*928, 946, 964*; **8**:*1032*; **10**:*1324, 1360, 1382, 1418*
mammals **1**:*88*; **2**:244; **3**:*312, 356, 374, 410*; **4**:*476*–477, *506, 550*; **5**:*592, 648, 664, 700*; **6**:*724, 844*; **7**:882, 898, 928, 946, 964, 965; **8**:*1032*–1033, *1110*; **10**:*1324*, 1325, *1360*, 1361, *1382*–1383, *1418*
calcium regulation **2**:268

cloaca **2**:185
ear ossicles **4**:555
embryo **7**:*1005*
estrogen cycle **2**:*269*
exocrine glands **2**:*274*
heart **1**:112; **4**:487
kidneys **2**:*186*
nonplacental **7**:882–883
placental **1**:*88*, 89; **2**:189, 245; **3**:*356, 410*; **4**:477, *506*, 550; **5**:*648*, 665, *700*; **6**:*724, 844*; **7**:899, 966; **8**:*1032*, 1033; **10**:1325, 1361, *1418*
reproductive structures **7**:994–*996*
respiratory systems
  gas exchange **8**:*1012*, 1019–1020
  ventilation **8**:1028
skeletal system **9**:1160, 1170
mammary glands **2**:*275*
  dolphin **2**:207
  human **2**:*275*; **4**:571
  mandrill **6**:*743*
  ungulate **2**:*275*
  whale **3**:394, *395*
Mammillaria
  hahniana **8**:1053
  longimamma **8**:1052
mammoth **2**:*244*, 245
manatee **2**:248; **5**:**700–717**
  Amazonian **5**:*700*, 703, *704*, 712, 717
  Antillean **5**:701
  buoyancy **5**:711
  circulatory and respiratory systems **5**:701, *711–712*
  communication **5**:710

digestive and excretory systems **5**:701, *713–714*
  diving **5**:712
  external anatomy **5**:701, *702–703*
  Florida **5**:701
  muscular system **5**:701, *708*
  nervous system **5**:701, *709–710*
  reproductive system **5**:701, *715–717*
  skeletal system **2**:250; **5**:701, *705–706*; **9**:1168
  tail **5**:*703*
  West African **5**:*700*, 701, 702, 704, 717
  West Indian **5**:*700*, 701, *702–703*, 704, 705, 706, 707, 709, 713, 715, 717
mandelonitrile **3**:340
mandibles **1**:135; **5**:*684*, 692, 693, 697
  of beetles **10**:1348
  of caterpillars **4**:499
  of centipedes **3**:339
  of weevils **10**:1359
mandibular arch **9**:1171
mandibular depressor muscle, crocodile **2**:156
mandibular gland **1**:30; **4**:532
mandrill **6**:**724–743**
  circulatory and respiratory systems **6**:725, *738*
  colors on face and buttocks **6**:725, *726*, 728, *730*, 738
  digestive and excretory systems **6**:725, *739–741*

external anatomy
6:725, 726–730
muscular system 6:725,
734–735
nervous system 6:725,
736–737
reproductive system
6:725, 742–743
skeletal system 6:725,
731–733
Mandrillus 6:724
mangabey 6:724, 725
mannose binding lectin
(MBL) 5:611
mantle
of giant clams 3:344,
346–347, 350
of octopuses 6:814
of oysters 3:348
marlin 8:1064, 1065
black, migration
8:1073
marmoset
brain 6:736
jaw 6:741
skull 6:732
marmot 9:1208, 1210
Marmota 9:1208
marrow
bone 1:108;
4:555–556; 5:618;
9:1168
stem cells 7:959
marsh grass 6:744–753
external anatomy
6:745, 746–748
internal anatomy
6:745, 749–750
reproduction 6:745,
751–753; 7:1001
Marsilea 2:280
marsupials 1:88; 4:551;
5:648, 700; 6:844;
7:885
cloaca 2:185
evolution 9:1193

marten 10:1324,
1325
American
10:1329–1330
beech 10:1330
pine 10:1329, 1339,
1340
reproduction 10:1341,
1343
Martes 10:1324
masseteric tubercle
9:1216
masseter muscles 6:853;
7:954, 955;
9:1216, 1217;
10:1334, 1428
Mastadenovirus 10:1314
mast cells 5:611, 615
mastication 2:173
Mastigophora 1:20–21;
2:164
mastodon 2:244
matamata 9:1198
mate-guarding, grizzly
bears 3:428
maxillae 5:684, 697
of centipedes 3:339
of lobsters 5:689
of spiders 9:1254
maxillipeds 5:685
of centipedes 3:333,
339, 340
of crabs 1:135, 139
of lobsters 5:689
mayfly, nymphs
8:1017
mazaalai 3:411
mealybug 1:33
mechanoreceptors
6:787, 788–789
earthworm 2:240
housefly 4:546
spider 6:791;
9:1259–1260
Mecoptera 4:538
medlar 1:36

medulla oblongata
(medulla) 6:796
of chimpanzees 1:98
of dolphins 2:199
of pumas 7:939
of trout 9:1287
of wolves
10:1394–1395
medusae 5:637, 638, 646,
647; 7:986–987
Megachilidae 4:526
Megachiroptera 3:292
Megadyptes antipodes
7:868
megagametocytes 7:1001
Megalapteryx
didinus 6:837
maximus 6:837
Megalonychidae 3:312;
9:1176
Megaloptera 4:538
Meganeuropsis 2:217
Megaptera novaeangliae
3:374
Megascolecidae 2:234
megaspores 7:1001
Megatherium 3:320;
9:1182
meiosis 1:67, 84, 87
and diatom
reproduction 2:171
melanophores 5:631;
6:789
melanophore-stimulating
hormone (MSH)
5:631
melatonin 2:273; 9:1202
Melinae 10:1324
Meliponinae 4:526
Meliursus ursinus 3:410
Melocactus 8:1053, 1056
melon (of cetaceans)
2:189, 193, 195,
201
membrane attack
complex 5:611

Mendel, Gregor 1:77–78
meninges 7:938
Menopon gallinae 5:693
menstruation 6:743;
7:999
meristoderm 8:1133
mermaids 5:717
Merostomata 9:1252
mesencephalon 10:1395
mesocuticle 9:1258
mesogleal cells 8:1089
mesoglea 3:354; 5:637,
638, 642; 8:1091
mesohyl 9:1162
mesophyll cells 7:924;
8:1145
mesophyll tissue 1:41,
42; 7:924
Mesothelae 9:1252, 1253
mesothorax 10:1348
Mespilus 1:36
metabolism, slow, of
sloths 9:1189
metacarpals 3:298;
10:1390, 1424,
1426
metamerism, in
earthworms 2:238
metamorphosis
in caterpillars 2:267
complete 4:505, 549;
5:699; 10:1345,
1358
in frogs 1:57, 62, 63;
2:267
in honeybees 4:537
incomplete 4:505;
5:699
metapleural gland 1:30
metapterygium 9:1243
metatarsals 10:1390,
1424, 1426
Metatheria 1:88; 5:648;
6:724, 844
metathorax 10:1348
Metazoa 8:1084–1085

metazoans **8**:*1084*–1085; **9**:1224

metencephalon **10**:1395

Metridiidae **8**:1085

Mexico
cacti **8**:*1050, 1051*
Montezuma pine **8**:1140
onza **7**:945
tortoises **9**:*1271*
wolves **10**:1385
woodpeckers **10**:*1405*

MHC *see* major histocompatibility complex

microbats **3**:292–293
echolocation **3**:305
tail **3**:*296*

Microchiroptera **3**:*292*, 293

Microcladia **8**:1131

microorganisms
excretion by **2**:187
plant- and cellulose-digesting **2**:181, 182
in elephants **2**:261
in rhinos **8**:1045
see also cellulose digestion

micropyle **8**:1147

microsporangia **7**:*1000*; **8**:1146

microtubules **1**:*75*
diatom **2**:169–170

microvilli **2**:179, *180*, 182, 183

Middle East
chameleons **5**:624
hyenas **5**:593

midge **4**:548
larvae **4**:547

midwives **2**:207

migration
billfish **8**:1073
crabs **1**:140

gray whales **3**:395
hummingbirds **5**:590
trout **9**:1291
wildebeest **10**:*1364*

mildew, powdery **6**:771

milk **2**:275
production **2**:271, *273*
seal **8**:1125
whale **2**:207

millipede **3**:*332*, 333; **8**:*1074*
defenses **3**:*340*
diplosegments **3**:335
exoskeleton **9**:1166
polydesmid **3**:340
porcupine–Velcro **3**:*340*

mimicry
of bees **4**:530
by caterpillars **4**:499

mineralocorticoids **2**:272

mink **10**:*1324*, 1325, *1331*, 1335
American **10**:1328–1329, 1341, 1343

miracidia **7**:987

mite **8**:*1074*, 1075; **9**:*1252*
on bees **4**:*531*

mitochondria **1**:*69*, 72–73; **5**:586; **6**:*767*; **9**:1188, 1290
of diatoms **2**:169
DNA **1**:83; **8**:1146
in muscle cells **7**:875; **8**:1067

mitosis **1**:25, 84–87; **7**:991

mitral valve **1**:*116*

moa **2**:223; **6**:837

Mollusca **3**:*342*; **6**:*812*–813

mollusks **3**:*342*; **6**:*812*–813
digestive system **2**:179

excretory system **2**:187
radula **2**:175, 179
shells **9**:*1164*–1165

molting
arthropods **9**:1167
caterpillars **4**:504–505
centipedes **3**:334
crabs **1**:141
dragonflies **2**:*211*,
insects **2**:267
birds
penguins **7**:871
woodpeckers **10**:1407

Monera (Prokaryotae) **1**:*46*

mongoose **5**:*592, 664*; **7**:*928*; **10**:*1324*

monkey
catarrhine monkeys **1**:89; **6**:725, 730
colobus
digestive system **6**:739
hands and feet **6**:*728*, 729
skull **6**:*732*
teeth **6**:741
drill **6**:725
dusky titi **6**:*729*
haplorrhine monkeys **6**:725
langur
digestive system **6**:739
jaw **6**:741
New World **1**:*88*; **4**:*550*; **6**:*724*, 725
cebids **6**:*729*, 730
vomeronasal organ **6**:737
Old World **1**:*88*, 89; **4**:*550*; **6**:*724*, 725
platyrrhine monkeys **1**:89; **6**:730

proboscis, digestive system **6**:739
red howler **6**:*729*
spider **6**:*734*, 735
black spider **6**:*729*
vervet **6**:*730*

monkey puzzle tree **8**:1139, *1140*, 1142

monocots **1**:*36*, 37; **6**:*744*, 824–825; **7**:*916*, 917; **8**:*1048*; **10**:*1300*
roots **6**:826–827
seeds **6**:*753*

Monocotyledonae **1**:*36*; **6**:*744*, *824*; **7**:*916*

monocytes **1**:*108*, 109; **5**:*609*

Monodontidae **2**:*188*

monoecious plants **7**:1002

Monognathidae **3**:*436*

Monoplacophora **3**:*342*; **6**:*812*

monospores **8**:1136

Monostroma **8**:1132

Monotremata **7**:*882*

monotremes **4**:551; **7**:*882*, 883, 885
body temperature **7**:*887*
brain **7**:891
cloaca **2**:185; **7**:883, *894, 895*–896
mammary glands **7**:*896*
skeletal system **7**:888–889

moose **7**:*967*

morel **6**:774

mosquito
larvae **4**:547
mouthparts **4**:*541*
sucking blood **4**:546

moss **2**:285

moth **4**:*494*, 495, *538*, 541
  apatelodod **4**:*494*
  bombycoid **4**:*494*, 495
  giant atlas **4**:495
  lappet **4**:*494*
  silkworm **4**:*494*, 504
  tent caterpillar **4**:*494*
  tobacco hornworm **4**:*496*
  vampire **4**:497
  see also hawkmoth
mother-of-pearl **3**:348
mouse **7**:*946*
  coprophagy **7**:961
  differs from rats **7**:*950–951*
  food and feeding **7**:961
  harvest **7**:*955*, 961, 963
  house, skull **7**:*953*
  New World **7**:*946*
  Old World **7**:*946*
  paws **7**:*949*
  wood **7**:961
Mucor **6**:771
Mucorales **6**:771
mucous membranes **5**:609
mucus **5**:609
mucus glands **2**:274
mud puppy **6**:801, 803
Mugiloidei **8**:*1060*
mule **10**:1437
mullet **8**:*1060*
multifidis cervicis muscle **2**:228
multiple fission **1**:24–25
Muntiacus **7**:*964*
Muridae **7**:*946*, 947
Murinae **7**:947
Muscidae **4**:*538*, 539
muscular system **6**:**754–769**
  antagonistic muscles **6**:*734*, *754*
  circular muscles **6**:*755*

force versus speed **6**:764
mode of action **6**:754, *756–761*
muscle cells (myocytes) **1**:66–67
  energy for **6**:761
red and white muscles **8**:1012, 1067; **9**:*1286*
smooth muscles **1**:67; **2**:254; **6**:755, *756*, *768–769*; **7**:890, 954; **10**:1392
striated/striped muscles **5**:630; **7**:890, 954; **10**:1392
  cardiac **1**:67, 115–116; **2**:254; **6**:755, *756*, *766–767*; **7**:*954*; **10**:1338
  fast- and slow-twitch **7**:973–974; **9**:1201
  skeletal **1**:66–67; **2**:254; **6**:755, *756–757*, *762–765*; **10**:1392
mushroom **6**:**770–779**
  external anatomy **6**:771, *772–775*
  internal anatomy **6**:771, *776*
  "magical" growth of **6**:776
  reproduction **6**:771, *777–779*
musk glands, crocodile **2**:152, 161
muskrat, tail **7**:955
mussel **3**:*342*, 350
  catch muscles **6**:768
  freshwater **3**:*342*, 355
Mustela **10**:*1324*, 1339

Mustelidae **3**:*410*; **6**:*844*, 845; **10**:*1324*, 1325, *1382*
mustelids **6**:*844*, 845; **10**:1325
Mustelinae **10**:*1324*, 1325, 1341
mutations **1**:85, 86–87
  in fruit flies **6**:793
mutualism, between giant clams and zooxanthellae **3**:352
mycelium, fungal **6**:*772–773*, 779
mycetome **5**:697
mycobacteria **5**:606
mycoplasmas **1**:48
mycorrhizae **6**:773, 831; **8**:1143
Myctophiformes **3**:439
myelencephalon *see* medulla oblongata
myelin sheath **4**:453–454; **8**:*1118*
Mygalomorphae **9**:*1252*
mygalomorphs **9**:*1252*
myiasis (sheep strike) **4**:548
Myidae **3**:*342*
Myliobatidae **9**:1239
Myliobatiformes **9**:*1238*, 1239
mylohyoid muscles **3**:*383*
myocytes (muscle cells) **1**:66–67
myofibrils **6**:757
myoglobin **2**:203; **6**:761, 839, 857; **7**:875, 876, 893, 973; **8**:1024, 1067, 1117; **9**:*1286*; **10**:1427
Myoida **3**:*342*
myomeres **8**:1102
  of hagfish **4**:452
  of sailfish **8**:*1066*

Myomorpha **7**:*946*, 947
myosepta **1**:127
myosin filaments **5**:672; **6**:754, *756–759*, *760*, *768*
myotomes **1**:127; **9**:1285, *1286*
Myriapoda **8**:*1074*
myriapods **3**:*332–333*
Myrmecocystus **1**:*26*
Myrmecophaga **3**:*312*
  tridactyla **3**:*312*
Myrmecophagidae **3**:*312*
Myrmicinae **1**:27
Mysticeti **2**:*188*; **3**:*374*
Myxine **4**:*446*, 448, 454
  glutinosa **4**:*446*
Myxini **4**:*446*, 448, 450
Myxinidae **4**:*446*, 448
Myxiniformes **4**:*446*, 448
Myxinikela siroka **4**:449
myxomatosis **4**:488
Myxomycota **8**:*1126*
Myxophaga **10**:*1344*

N

narwhal **2**:*188*, 190
  jousting **2**:206
natural selection **1**:86
Nautiloidea **6**:*812*, 813
nautiloids **6**:*812*, 813
  buoyancy **6**:815
Nautilus **6**:813
Navicula **2**:*166*, *167*, *168*
navigation, baleen whales **3**:385
Neanderthal people **4**:*550*
necrotrophs, fungi as **6**:774
nectar **1**:39; **10**:1308
nectaries **1**:39
Needham's sac **6**:823
Nemalion multifidum **8**:1132

Nemamyxine **4**:*446*, 448
Nematocera **4**:*538*, 539
nematocysts (cnidae)
    5:637, *640*; **8**:1086,
    1092–1093
nematode (roundworm)
    6:*774*
Neobalaenidae **3**:*374*
Neobatrachia **1**:*54*, 55
Neomyxine **4**:*446*, 448
neoteny **6**:803, 811
Nepenthaceae **10**:1301
Nepenthales **10**:*1300*,
    1301
Nepenthes **10**:1304–1305
    kinabaluensis **10**:1303
    rajah **10**:1301, 1308
nephridia **2**:187
    of earthworms **2**:237,
    242
nephrocytes **9**:1263
Nephropidae **5**:*684*
Nephropoidea **5**:*684*
nerve cells *see* neurons
nerve net **6**:792
    of jellyfish **5**:637, 641
    of sea anemones
    8:*1090*
nervous system
    6:**780–799**
    central (CNS) **4**:*562*;
    6:*780*, 794,
    796–799; **7**:938,
    956; **9**:1259
    and the endocrine
    system **2**:264; **4**:563
    invertebrate **6**:*792*–793
    mode of action **6**:781,
    782–785
    peripheral (PNS)
    4:562; **6**:*780*, 794;
    7:938, *956*
    autonomic nervous
    system (ANS)
    4:563; **6**:*766*, 794,
    795; **7**:938

of hummingbirds
    5:587
somatic nervous
    system **4**:563;
    6:794–795
sensory and motor
    systems **6**:781,
    786–791
somatic **4**:563
vertebrate **6**:*794–799*
development in the
    embryo
    7:*1004–1005*
Nesolagus **4**:*476*
neural plates **9**:1272
neural spines **9**:*1172*
neurohormones **2**:270,
    275
neuromasts **3**:442;
    9:*1247*, *1287*,
    1288
neuromuscular junctions
    6:760–*761*, 791
neurons **1**:66; **3**:366;
    4:502; **6**:768, 781,
    782–785; **7**:909
    giant, of cephalopods
    6:783, *820*
    named **6**:793
    types
        interneurons **2**:256;
        4:562; **6**:782, *783*;
        8:1042
        motor **2**:256; **4**:562;
        6:782, *783*; **7**:*956*;
        8:1042
        sensory **2**:256;
        4:562; **6**:782, *783*;
        8:1042
Neuroptera **4**:*538*
neurotransmitters **4**:562;
    6:*769*
    see also acetylcholine
neutrophils **1**:108, 109;
    5:608, *610*, 611

newt **6**:**800–811**
    circulatory and
    respiratory systems
    6:801, *808–809*
    crested **6**:*802*–803,
    804, *805*, 807, *809*,
    810, 811
    external anatomy
    6:801, *802–804*
    larvae **6**:804
    nervous system **6**:801
    red-spotted **6**:*810*
    regrowing damaged
    parts **6**:803
    reproductive system
    6:801, *810–811*
    ribbed **6**:*806*
    skeletal and muscular
    systems **6**:*805–806*
    walking on land **6**:805
New Zealand, kiwis
    6:834
Nicaragua, sloths **9**:*1183*,
    1193
Nicotiana
    glauca **7**:*916*
    tabacum **7**:925
nicotine **7**:925
nictating membrane
    2:224; **4**:*471*; **5**:703
night crawler **2**:*234*, 235
    digestive tract **2**:242
    external anatomy
    2:*236*
    gaseous exchange
    2:241
    internal anatomy
    2:*238*, 239
    nervous system **2**:240
    reproductive system
    2:243
nociceptors **6**:789
nodes and internodes
    6:827; **7**:918
norepinephrine **1**:121;
    2:272

North America
    grizzly bears **3**:411
    ground squirrels
    9:*1212*
    pikas **4**:480
    wolverine **10**:*1330*,
    1331
North Carolina, Venus
    flytrap **10**:1301
Notacanthiformes **3**:*436*
notochord **4**:476;
    7:1004; **9**:1158,
    1172
    of hagfish **4**:*448*, *452*;
    9:*1159*
Notomyotida **9**:*1224*
Notomyxine **4**:*446*, 448
nucleus **1**:70
numbat **3**:321
nyala **10**:*1376*
Nyctimene spp. **3**:*292*

O

Obelia **5**:637
occipital condyles **9**:1173
ocean depths
    fish **3**:439, 440
    gulper eels **3**:*437*, 439
ocelli
    of ants **1**:31
    of honeybees **4**:534
    of jellyfish **5**:643
    of myriapods **3**:337
    of starfish **9**:1232
ocelot **7**:928
Ochotona **4**:*476*, 480
Ochotonidae **4**:*476*
Octopoda **6**:*812*
Octopodiformes
    6:*812*
octopods **6**:*812*
octopus **6**:**812–823**
    blue-ringed **6**:822
    circulatory and
    respiratory systems
    6:813, *821*

common **6**:*815–816,*
819, 821, 822, 823
mouth **2**:*175*
digestive and excretory
systems **6**:813, *822*
external anatomy
6:813, 814–*817*
eye **6**:*788,* 817, *820*
giant **6**:*818*
ink **6**:*817*
jet propulsion **6**:818
muscular system **6**:813,
818
nervous system **6**:792,
813, *819–820*
reproductive system
6:813, *823*
tentacles **9**:*1165*
Odobenidae **8**:*1110*
Odocoieus **7**:*964*
Odonata **2**:*208;* **5**:*692*
Odontoceti **2**:*188;* **3**:*374*
Odontomachus **1**:*26*
okapi **3**:*356,* 357, 359,
360
Okapi johnstoni **3**:*356*
olecranon process **8**:*1039*
olfactory epithelium
1:*16;* **3**:*304–305*
Oligochaeta **2**:*234*
oligochaete worms **2**:234
olm **6**:801
omasum **7**:980; **10**:*1375,*
1376
ommatidia **4**:534
of dragonflies **2**:215
of houseflies
4:545–546
omnivores, teeth **2**:174
Oncorhynchus
clarkii **9**:*1292*
gilae **9**:*1292*
mykiss **9**:*1280,* 1284
onza **7**:945
oocytes, primary **7**:992
oomycota **6**:770

operculum **1**:60
Ophioglossaceae **2**:*276*
Ophioglossales **2**:*276*
Ophioglossum **2**:*276*
Ophiuroidea **9**:*1224*
Ophrys **6**:832
Opiliones **8**:*1074,* 1075
opisthosoma (abdomen)
8:1076; **9**:*1252,*
1254, 1257, 1258
opsonization **5**:611, 613
Opuntia **8**:1049, 1051,
1058
Opuntioideae **8**:*1048,*
1049, 1052
oral (buccal) cavity
2:173
orangutan **1**:*88,* 89, 97,
103; **4**:*550*
orbicularis oris muscle
6:*764*
orca *see* whale, killer
orchid **6**:*744,* **824–833**
bee **6**:832
bog **6**:832
epiphytic **6**:825, 826,
827, 830, *831*
external anatomy
6:825, *826–828*
grass pink **6**:825, *827,*
832, 833
internal anatomy
6:825, *829–831*
reproduction **6**:825,
832–833
slipper **6**:832
underground **6**:827,
828
Orchidaceae **6**:*744,* 824,
825
Orchidales **6**:*824,* 825
Orchidoideae **6**:*824,*
825, 826
Orchis **6**:825
organelles **1**:*64,* 69,
70–74

of diatoms **2**:168–169
evolution from
prokaryotes **1**:69;
2:170
Ornithorhynchidae
7:*882,* 883
Ornithorhynchus anatinus
7:*882*
Orthomyxoviridae
10:1314
Orthopoxvirus **10**:1314
Oryctolagus cuniculus
4:*476*
Oryza **6**:752
Osmeridae **9**:1281
osprey **2**:*220*
ossicles
ear **4**:555, *565*
scleral **9**:1199
of a starfish skeleton
9:1228
ossicones **3**:359–360, *363*
Osteichthyes **1**:*122,* 123;
3:*436,* 437; **4**:*460;*
8:1061, *1096,* 1097;
9:*1238, 1280,* 1281
osteoblasts **9**:*1168*
osteoclasts **9**:1168
osteoderms **2**:151, 153
Osteolaemus **2**:*148*
osteoporosis **9**:1168
ostia **5**:688; **10**:1355
ostrich **6**:**834–843**
circulatory and
respiratory systems
6:835, *841*
digestive and excretory
systems **6**:835, *842*
external anatomy
6:835, *836–837*
extinct relatives **6**:837
keeping cool **6**:841
muscular system **6**:835,
839
nervous system **6**:835,
840

reproductive system
6:*835, 843*
skeletal system **6**:835,
838
hind limb bones
9:*1175*
two species of? **6**:837
Otariidae **8**:*1110*
otter **6**:**844–861;**
10:*1324*
ancestry **6**:847
Asian short-clawed
6:*849,* 852, 858
circulatory and
respiratory systems
6:845, *856–857*
clawless **6**:*849,* 851,
852, 855, 858, 861
digestive and excretory
systems **6**:845,
858–859
Eurasian river **6**:*844,*
847, 848, 859,
860–861
external anatomy
6:845, *846–849*
eyes **6**:848–849, 853,
855
fur **6**:847–848, 849,
850
giant **6**:*844,* 845, *847,*
848
Lontra-type **6**:*844,* 845
Lutra-type **6**:*844,* 845
muscular system **6**:845,
853
nervous system **6**:845,
854–855
North American river
6:*844,* 845, *848,*
861
number of species
6:845
reproductive system
6:845, *860–861*
scent **6**:854, 860

otter *(continued)*
　sea **6**:*844*, 845,
　　846, 847, *849*,
　　851–861
　sizes **6**:*849*
　skeletal system **6**:845,
　　850–852
　sociable or solitary
　　6:861
　streamlining **6**:848
ova (eggs) **7**:*990*, 999
　development in
　　humans **7**:*995*
　see also eggs
ovaries
　albatross **1**:19
　apple tree **1**:*43*
　chameleon **5**:634
　eagle **2**:232
　mammalian **7**:995
　ovarioles **10**:1357
　oviducts **7**:996
　ovipositor **1**:35
　　of dragonflies
　　　2:213, 218
　　as a stinger **1**:35;
　　　4:533
　　of weevils **10**:1359
ovoviviparity **3**:409
　in coelacanths **1**:131
　in sheep botflies
　　4:549
ovulation **7**:*995*
　delayed **10**:1342
ovules **8**:1146–1147
ox, Vu Quang **3**:360
oxpecker **8**:*1035*
oxygen **8**:1012–1013,
　1015
oxytocin **2**:271
oyster **3**:*342*, 345, 355
　American **3**:349

P

pacemaker
　in decapods **6**:767
　in vertebrates **6**:767

Pacific Ocean
　gray whales **3**:375
　sea horses **8**:*1098*
padloper, speckled
　9:*1279*
Paenungulata **2**:*244*
Pakicetus **2**:192
Palaeudyptes antarcticus
　7:*872*
palate, human **2**:*173*
palea **6**:751, *752*, 753
Paleognathae **6**:*834*
Paleoheterodonta **3**:*342*
paleopallium **4**:564–565
Palinuroidea **5**:*684*, 685
pallial muscles, of clams
　3:*350*
palolo worms **7**:988
palpal organs **9**:*1264*,
　1265
Pan **4**:*550*
　paniscus **1**:*88*
　troglodytes **1**:*88*
pancreas
　digestive function
　　2:184; **5**:589; **6**:741;
　　8:1107
　endocrine function and
　　hormones **1**:70;
　　2:272, 273; **4**:571;
　　8:1107
panda, giant **3**:*410*, 411,
　419
Pandinus **8**:1075
Pandionidae **2**:*220*
pangolin **3**:315; **9**:1185
　Cape **9**:*1181*
panicles **6**:751
Panizza, foramen of
　2:159–160
panther, Florida **7**:944
Panthera **5**:665; **7**:932
　leo **5**:*664*, 665
　onca **5**:*664*
　pardus **5**:*664*
　tigris **5**:*664*
Papilionoidea **4**:*494*, 495

Papillomavirus **10**:1314
Papio **6**:*724*
Papovaviridae **10**:1314
Papua New Guinea,
　echidnas **7**:883
papulae, of starfish
　9:*1233*
parabronchi *see* air sacs
Parahyaena brunnea **5**:*592*
paramecium **1**:21, 22, *25*,
　75; **2**:167
paranthropines **4**:*550*
Paranthropus **4**:*550*
parapodia, of polychaetes
　2:234
parasites
　on bovids **10**:1374
　lampreys **4**:447
　lice **5**:692, 693
　ticks **5**:698
　viroids **10**:1317
parathyroid gland **2**:268
　of eagles **2**:233
parenchyma cells **1**:67;
　6:829, *830*, 831;
　7:*924*, 925;
　8:*1055*–1056, 1144;
　10:1306
parental care
　centipedes **3**:341
　frogs and toads **1**:63
　octopuses **6**:823
　scorpions **8**:*1083*
　sea horses **8**:1109
parenthosome **6**:776
parietal bone, in
　chameleons **5**:628,
　　629
parietal organ **1**:60
Parkinson's disease **4**:565
parthenogenesis **7**:*986*,
　988–*989*
patagium **9**:*1213*, 1217
pathogens **5**:606
Pauropoda **3**:*332*
pauropods **3**:*332*, 333
　legs **3**:*335*

Paxillosida **9**:*1224*
peach **1**:37, 44
pear **1**:37
　European **1**:*36*
pearls **3**:*348*
peccary **4**:507
Pecora **3**:*356*
pectines **8**:1077, 1082
pectoral girdle **9**:1174
　of otters **6**:852
　of rays **9**:*1242*, 1243
pectoralis major muscles
　1:15; **5**:586; **6**:*839*;
　7:875
pectoralis minor muscles
　1:15
pectoralis muscle **4**:561
　of bats **3**:302
　of birds **1**:15; **3**:302;
　　10:1412
pedal glands **8**:1047;
　10:1363, 1373
pedal laceration **8**:1094,
　1095
pedicellariae **9**:*1227*,
　1231
pedicels **7**:*970*, 971, 972;
　9:1257
Pediculidae **5**:693
Pediculus humanus **5**:*692*,
　693
pedipalps **9**:1253
　scorpion **8**:1075, *1076*,
　　1077, *1078*
　spider **2**:175; **9**:*1254*,
　　1262
Pelecanoididae **1**:*10*
pelvis (pelvic girdle)
　9:1174
　frog **9**:1159–1160
　otter **6**:852
　vestigial **9**:*1175*
　zebra **10**:1426
penguin **3**:379;
　7:**868–881**
　Adélie **7**:*868*, 870, 873,
　　874, 880

blue **7**:*868*, 870
chinstrap **7**:*868*, 873
circulatory and
    respiratory systems
    7:869, *877–878*
crested **7**:*868*
digestive and excretory
    systems **7**:869, *879*
diving by **7**:878
emperor **7**:869, 870,
    871, 872, 873, 876,
    877, *879, 881*
evolution **1**:14; **7**:874
external anatomy
    7:869, *870–872*
feathers **7**:*871–872,
    875, 881*
Fiordland crested **7**:880
Galápagos **7**:*870*, 877
gentoo **7**:*868*, 873, 876
Humboldt **7**:877
king **7**:869, 870, 876,
    880, 881
Magellanic **7**:881
muscular system **7**:869,
    875
nervous system **7**:869,
    876
origins **7**:*872*
reproductive system
    7:869, *880–881*
salt glands **7**:879
skeletal system **7**:869,
    873–874
swimming and
    porpoising
    7:*870–871,
    873–874*
in the tropics **7**:877
yellow-eyed **7**:*868,
    869*
Penicillium **6**:777
penis
of armadillos **3**:323
of echidnas **7**:895
of humans **4**:*572*;
    7:995

of hyenas **5**:605
of platypuses **7**:*895*
of pumas **7**:*944–945*
of reptiles **9**:1206
of rhinos **8**:1047
of whales **3**:383, *394*
Pennales **2**:*164*
pentadactyl limbs **5**:629
Pentalagus furnessi **4**:*476*
peptidoglycan **1**:48–49
perch **8**:*1060*, 1061;
    9:*1280*
scales **9**:*1283*
Perciformes **8**:*1060*,
    1061; **9**:*1280*
Pereskia diaz **8**:*1052*
Pereskioideae **8**:*1048*,
    1049
pericardium **1**:113
pericarpel **8**:1059
pericycle **1**:*65*; **7**:*925*;
    10:*1307*
peridermis **8**:1145
periosteum **7**:*971–972*
Perissodactyla **8**:*1032*,
    1033; **10**:*1360*,
    1361, 1366, *1418*
peristalsis **2**:178; **3**:*371*
peristomium, of
    earthworms **2**:234,
    236, 240, 242
peritrophic envelope
    2:214
peroxisomes **1**:73
Peru, weevils **10**:*1345*
Petauristinae **9**:*1208*
petiole **10**:1302
petrel **1**:*10*, 11
see also diving petrel;
    storm petrel
Petrogale **5**:649
Petromyzontidae **4**:447
Petromyzontiformes
    4:447
petrosal bones **7**:888
Petunia **7**:*916*
Peyote **8**:1052

Phacochoerinae **4**:*506*
Phaeophyceae **8**:*1126*,
    1127
Phaeophyta **8**:*1126*,
    1127
Phaethornithinae **5**:*580*
phagocytosis **1**:23, *66*
phagosomes **5**:611
Phalaenopsis **6**:*825*
Phalangerida **5**:*648*, 649
Phalangeridae **5**:*648*
phalanges
    bat **3**:298
    chameleon **5**:*629*
    giant anteater **3**:321
    giraffe **3**:363
    human **4**:558
    seal **8**:1116
    whale **2**:196
Phallus impudicus **6**:*779*
pharyngeal spines, turtle
    2:*176*
pharyngeal teeth,
    sandworm **2**:178
pharyngeal
    ultimobranchial
    glands **2**:268
pharynx **2**:176
phenotypes **1**:77
    wild-type **1**:78
pheromones
    of lions **5**:676
    as sexual attractants
        2:274
    crab **1**:140
    elephant **2**:263
    porcupine **7**:908
    scorpion **8**:1082
    social insect **2**:275
    ants **1**:30, 31
    as territory markers
        2:274
    triggering ovulation,
        porcupines **7**:914
phloem **1**:40, *65*, 67;
    6:829; **7**:924, 925;
    8:*1144*, 1145

Phocidae **6**:*844*;
    8:*1110*
Phocoenidae **2**:*188*
Phoebetria **1**:11
    fusca **1**:*10*
phoresy **8**:1075
Phoridae **4**:*538*
photocytes **10**:1349
photoreceptors **6**:789
    extraretinal **9**:1202
photosynthesis **1**:74;
    8:1013; **10**:1307
    C3 and C4 **6**:750;
        8:1031
Phthiraptera **5**:*692*
Phthiridae **5**:693
Phthirus pubis **5**:693, *695*
phycocyanin **8**:1127
phycoerythrin **8**:1127,
    1131
phyllodes **10**:1302
physa **8**:1086
Physeteridae **2**:*188*
Phytophthora infestans
    6:770
Picidae **10**:*1404*, 1405
Piciformes **10**:*1404*
Picinae **10**:*1404*
Picoides **10**:1405
Picornaviridae **10**:1314
picornavirus **10**:*1321*
piculet **10**:*1404*, 1405,
    1408, 1409, 1411
    olivaceous **10**:*1409,
    1411*
Picumninae **10**:*1404*,
    1405
pig **3**:*356*; **4**:507; **7**:*964*;
    10:*1360*
pigeon
    digestive system **2**:*178*
    magnetic-field
        detection **6**:791
pika **4**:*476*, 477, 479,
    480
    skull **4**:*483*
    steppe **4**:*479*

pike **9**:*1280*

pili *see* fimbriae

pill bug (wood louse) **5**:692

Pilobolus **6**:778–779

Pilosa **3**:*312*; **9**:*1176*

Pinaceae **8**:*1138*, 1139

pine **8**:*1138*
  bristlecone **8**:1139, 1145
  cones **7**:*1000*, 1003; **8**:*1146, 1147*
  Monterey **8**:1148
  Wollemi **8**:1139

pineal gland (pineal body; parietal eye; third eye) **1**:60; **2**:273; **9**:*1202, 1274*

Pinnipedia **3**:*410*; **8**:*1110*; **10**:*1382*

pinnipeds **8**:1111
  origins **8**:*1113*

pinocytosis **1**:*66*

pipefish **8**:*1096*, 1109

pistils **1**:43; **6**:752

pitcher plant **10**:1301, *1303*, 1307, 1311
  American **10**:*1300*, 1305
  northern **10**:*1302*
  swamp **10**:1303, 1304
  Western Australian **10**:*1300, 1304*

pituitary gland **4**:*571*
  hormones **2**:*266*, 268, 270–271; **7**:*997*–998

placenta (afterbirth) **3**:330, 356; **7**:*1005*; **10**:1379
  eating the **10**:1381
  hemochorial **6**:743

planarians, excretion **2**:*187*

Plantae **6**:*744, 824*; **7**:*916*; **8**:*1048*, *1138*

plantigrade stance
  bears **3**:*418*
  hares **4**:483
  humans **4**:553
  monkeys **6**:733
  otters **6**:852
  rats **7**:953
  squirrels **9**:1214

plants **2**:276; **6**:*744, 824*; **7**:*916*; **8**:*1048, 1138*; **10**:*1300*
  alternation of generations **7**:*993*
  carnivorous **10**:1300–1301
  cells **1**:67; **6**:*829*
  crassulacean acid metabolism (CAM) **6**:830; **8**:*1057*
  flowering **7**:917
  growth hormones **2**:266
  lower vascular **1**:*36*; **2**:276–277; **6**:*744, 824*; **7**:*916*; **8**:*1048*; **10**:*1300*
  nonvascular **1**:*36*; **6**:*744, 824*; **8**:*1048*; **10**:*1300*
  poisons **2**:187
    curare **6**:761
    in the potato family **7**:*925*
  polyploid **7**:927
  reproduction, asexual **7**:*988*, 989
  seed-bearing **1**:*36*, 37; **6**:*744, 824*; **7**:*916*, 917; **8**:*1048, 1138*; **10**:*1300*
  support structures **9**:*1161*
  vascular **1**:*36*–37; **2**:276; **6**:*744, 824*; **7**:*916*–917; **8**:*1048, 1138*; **10**:*1300*

plasma **1**:108; **5**:618

plasma membrane *see* cell membrane

plasmids **1**:*51*, 53, 77, *81*

plasmodia **8**:*1126*

Plasmodium falciparum **5**:*607*

plasmogamy **6**:777

plastron **8**:1018; **9**:*1198*, 1268–1269

Platanistidae **2**:*188*

platelets **1**:108, *109, 119*

Platybelodon **2**:247

Platycerium **2**:*276*

platypus **7**:**882–897**
  ancestors **7**:886
  bill **7**:*886*–887, 889, *892, 894*
  circulatory and respiratory systems **7**:883, *893*
  digestive and excretory systems **7**:883, *894*
  diving by **7**:893
  electroreceptors **6**:*790*
  external anatomy **7**:*883, 884–887*
  first discovered **7**:885
  muscular system **7**:883, *890*
  nervous system **7**:883, *891–892*
  reproductive system **7**:883, *895–897*
  skeletal system **7**:883, *888–889*
  venomous spurs **7**:*887*, 890

Platyrrhini **1**:*88*; **4**:*550*; **6**:*724*

Platysternidae **9**:*1194*, *1266*

play
  dolphin **2**:*207*
  wolf **10**:*1403*

Plectomycetes **6**:*770*, 771

pleopods **1**:135

plasma membrane *see* cell membrane

Plethodontidae **6**:*800*, *801*

Pleurodira **9**:*1194*, *1266*

Pleurosigma **2**:166

plum **1**:37, 44

pneumatocysts **8**:*1131*

pneumonia **5**:607

Pogonomyrmex **1**:*26*, 27

poison glands *see* venom glands and venom

polecat **10**:*1324*, 1325, 1328
  African (zorilla) **10**:1330–1331, 1336
  European **10**:1335, *1336, 1343*
  skull **10**:*1333*
  feet **10**:*1329*
  marbled **10**:*1330*, 1336, 1341
  reproduction **10**:1341
  thin **10**:1330

Polemoniales **7**:*916*

pollen **2**:276–277; **7**:1001, 1002

pollen-baskets **4**:529

pollination
  of apple trees **1**:43, *44*
  by bats **7**:*1002*; **8**:*1058*
  of cacti **8**:*1058*
  of giant sequoias **8**:1146–1147
  of grasses **6**:751
  of orchids **6**:832, *833*
  of Venus flytraps **1**:131

Polychaeta **2**:*234*

polychaete worms **2**:234, *235*
  gills **2**:241
  mouthparts **2**:236
  and reproduction **7**:988

polyembryony **7**:986

polygamy, squirrels **9**:1222

Polyphaga **10**:*1344*, 1345

Polyplacophora 3:*342*; 6:*812*
polyploid cells 1:76; 7:*927*
Polypodiaceae 2:*276*
polyps 5:637; 8:1087
  coral 8:1087
  hydrozoan 7:986
  jellyfish 5:638, 646; 7:986, *987*
Polysiphonia 8:*1134*
pome 1:*36*, 44
Pomoideae 1:*36*
Ponerinae 1:27
Pongo 4:*550*
  abelii 1:*88*
  pygmaeus 1:*88*
Pontoporidae 2:*188*
porcupine 7:**898–915**
  African 7:*902, 905, 907*, 914–915
  Brazilian 7:*906*
  brush-tailed 7:*900*, 902
  circulatory and respiratory systems 7:899, *910–911*
  crested 7:*900*, 904–*905*
  digestive and excretory systems 7:899, *912–913*
  external anatomy 7:899, *900–903*
  hairy dwarf 7:*902*, 912
  Indonesian 7:*902*
  long-tailed 7:*900*, 902
  muscular system 7:899, *906–907*
  nervous system 7:899, *908–909*
  New World 7:*898*, 901, *902–903*
  North American 7:*900–906*, 908, 910, *912–915*
    craving for salt 7:913
  Old World 7:*898*, 901, 912

pheromones 7:908, 914
prehensile-tailed 7:*899*, 912
  reproductive system 7:899, *914–915*
  skeletal system 7:899, *904–905*
  South American tree 7:*903*
  spines (quills) and rattles 7:899, *900–901, 907*
  stump-tailed 7:*903*
  three crested 7:*901*
Porphyra 8:*1126*, 1127, 1134, 1135–1136, 1137
porpoise 2:*188*, 190
  harbor, skull 2:*195*
  teeth 2:*195*, 204
porpoising 2:*192*; 8:*1031*
Portuguese man-of-war 5:637, *640*
Portunidae 1:*132*
possum 5:*648*, 649
Postelsia palmaeformis 8:1131
potato 7:**916–927**
  eyes 7:*920, 921*, 923
  flowers 7:*921, 926, 927*
  fruit 7:*922, 926*, 927
  reproduction 7:917
    asexual 7:926, 989
    sexual 7:926–927
  rhizomes 7:918, *919*, 920–921, 989
  roots 7:918–920, 924–925
  "seed" 7:*918, 919*
  stems and leaves 7:918, *919*, 923–*924*
  sweet 7:917, *922*
  tubers 7:918, *919*, 920, 921–*922, 923, 925*, 926
  varieties 7:926
potato blight 6:770

potato family 7:*916*, 917
  poisons in 7:*925*
Potoroidae 5:649
potoroo 5:649
Poxviridae 10:1314
poxvirus 10:1322
prairie dog 8:1026; 9:*1208*
precocious young 3:331, 373; 4:493; 6:843
preen gland (uropygial gland) 1:13; 10:1409
preorbital glands 10:1363
primates 1:88, 89; 4:*550*, 551; 6:*724–725*
  ancestors 6:727
  diets 1:101
  heart 6:738
  limbs 6:733
  locomotion 6:734–735
  manual dexterity 6:733
  reproductive biology 6:742
  senses 6:737
    color vision and frugivory 6:740
    stereoscopic vision 6:*736*
  varied lifestyles 6:729
Primates 1:*88*; 4:*550*; 6:*724*
primordia 6:774
prion 10:*1318*
Priscacara liops 9:*1158*
Pristiformes 9:*1238*
Pristina leidyi 2:243
Proboscidea 2:*244*
proboscideans 2:247
proboscis
  of butterflies and moths 4:*497*
  of houseflies 4:541
Procellariidae 1:*10*
Procellariiformes 1:*10*
processus entoglossus 5:*630*, 633

processus spinalis 1:93; 6:732
Procoptodon goliah 5:652
proctodaeum 1:18; 2:162
procuticle, dragonfly 2:210
Procyonidae 3:*410*
progesterone 2:269; 7:*995*, 999
Prokaryotae (Monera) 1:*46*
prokaryotes 1:20, *46*, 50, 64; 2:170
  mitosis in 1:*85*
prolactin 2:271, *273*
promiscuity 4:573
pronghorn 3:*356*; 7:*964*; 10:*1360, 1371*
pronking 10:1371
propterygium 9:1243
Proscyllidae 4:*460*
prosimians 6:737
prosoma (cephalothorax) 1:134; 8:*1076*; 9:1252–1253, *1257*
prostate gland 7:995
prostomium, of earthworms 2:240
Proteidae 6:801
proteins 7:984
  carrier 1:52
  manufacture in the cell 1:70, 71, 81–83
Proteles cristatus 5:*592*
Protelinae 5:*592*
Proteobacteria 1:*46*, 47
prothoracic glands, arthropod 2:267
prothorax 10:1348
prothrombin 1:119
Protista 2:*164*; 8:*1126*
protists 1:20; 2:164–165
protocorm 6:*833*
Protopteridae 1:*122*
Protosciurus 9:1215
protozoans 1:20
  amoeboid 1:*20*, 21

protozoans *(continued)*
  ciliated **1**:*20*, 21
  flagellated **1**:*20*
  parasitic **1**:*20*, 21
proventriculum **7**:879
proventriculus **1**:18, 32;
    2:230; **5**:589;
    10:1415
Prunus **1**:*36*, 38, 44
prunus **1**:*36*
pseudobulbs **6**:827, 828
pseudopenis, of hyenas
    5:602, 603, *604*
pseudopodia **1**:*21*, 22
pseudoscorpion **8**:*1074*,
    1075
Pseudoscorpiones
    8:*1074*, 1075
Pseudotriakidae **4**:*460*
Psilophyta **2**:*276*
Pteridium **2**:*276*, 279
Pteridophyta **2**:*276*
Pteriomorpha **3**:*342*
Pterocactus **8**:1049
Pteromyinae **9**:1209
Pteronura brasiliensis
    6:*844*
Pteropodinae **3**:*292*
Pteropus spp. **3**:*292*, 293
pterosaur, wings **3**:*299*
pterostigma **2**:211–212
pterygoid bones **3**:319
pterygoid muscles
    9:1217
ptilinal suture **4**:540
Puccinia **6**:771
pudu **7**:966
puffball **6**:771
  giant **6**:778
  wolf-fart **6**:*778*
pufferfish, venom
    6:783
pulmonary artery **1**:113;
    4:567
pulmonary valve **1**:*113*
pulmonary veins **1**:115
pulse **1**:117

puma **7**:**928–945**
  circulatory and
    respiratory systems
    7:929, *940–941*
  digestive and excretory
    systems **7**:929,
    942–943
  external anatomy
    7:929, *930–933*
  killing prey **7**:*943*
  muscular system **7**:929,
    936–937
  nervous system **7**:929,
    938–939
  power and agility
    7:*936*, *937*
  reproductive system
    7:929, *944–945*
  skeletal system **7**:929,
    934–935
  whiskers **7**:*933*
pupae **4**:526
  butterfly and moth
    4:*504*, *505*
  honeybee **4**:537
  housefly **4**:*549*
pupils, of cats **7**:*939*
Purkinje nerve cells
    6:797
pus **5**:608, 611
Pycnogonida **9**:*1252*
Pycnopodia helianthoides
    9:*1224*
pygidium, of earthworms
    2:234, 236
Pygoscelis **7**:*868*, 873, 876
pygostyle **6**:838; **7**:873;
    10:1410, 1411
pyloric ceca **9**:1291
pyloric sphincter
    (pyloric valve)
    9:1220, 1277
pylorus **2**:183
Pyracantha **1**:*36*
Pyrenomycetes **6**:770,
    771
Pyrrophyta **8**:*1126*

Pyrus **1**:37
  communis **1**:*36*
pyruvate (pyruvic acid)
    1:*72*, 73
python **3**:*396*, *400*
  heat-detection **3**:406
  keeping eggs warm
    3:405
Pythonidae **3**:*396*
Pyxis **9**:1268

Q
quillwort **2**:284
quince **1**:*36*
quokka **5**:649, 653, 656

R
rabbit **4**:*476*
  Annam striped **4**:*476*
  bristly **4**:*476*
  bushman **4**:*476*
  Central African **4**:*476*
  European **4**:*476*, 479
  pygmy **4**:*476*
  red **4**:*476*
  Ryuku **4**:*476*
  short-eared **4**:*476*
  skull **4**:*483*
  Sumatran short-eared
    4:*476*
  volcano **4**:*476*
  young **4**:493
raccoon **3**:*410*; **5**:*592*,
    664; **7**:*928*;
    10:*1324*
Radiolaria **2**:*164*
radiolarians **1**:*21*; **2**:*164*,
    165
  cell wall **2**:167
radioulna **1**:58
radius
  albatross **1**:*14*
  eagle **2**:226
  seal **8**:1116
radula **2**:175, 179
  octopus **2**:*175*; **6**:*817*,
    822

ragworm **9**:1163
rainbow plant **10**:1301,
    1303
Rajiformes **9**:*1238*
Ramphomicron **5**:582–583
ram's horn shells **6**:813
ram ventilation **8**:1027,
    1071
Rana **1**:55
  berlandieri **1**:*54*
  catesbeiana **1**:*54*
  esculenta **1**:*54*
  temporaria **1**:*54*
Rangifer tarandus **7**:*964*
Ranidae **1**:*54*, 55, *56*
ranids **1**:55
raphe, of diatoms
    2:166
Raphidoptera **4**:*538*
raspberry **1**:37, 38
rat **7**:**946–963**
  ancestors **7**:949
  black **7**:950
  brown **7**:*947–963*
  circulatory and
    respiratory systems
    7:947, *958–959*
  communication **7**:956
  dentition **7**:*950*, 953
  digestive and excretory
    systems **7**:947,
    960–961
  external anatomy
    7:947, *948–951*
  kangaroo **7**:*951*
  muscular system **7**:947,
    954–955
  nervous system **7**:947,
    956–957
  New World **7**:*946*
  Old World **7**:*946*
  poisons and **7**:960
  pouched **7**:*950*, *961*
  reproductive system
    7:947, *962–963*
  skeletal system **7**:947,
    952–953

used for
  experimentation
  7:957
ratites **6**:*834*
rat-kangaroo **5**:649
rat-tailed maggot **8**:*1017*
rattlesnake **3**:401
  heat-detection **3**:406
Rattus **7**:947
ray **4**:461; **9**:*1238*, 1239
  blue-spotted **9**:*1245*
  buoyancy **9**:1249
  butterfly **9**:1251
  cow-nose **9**:1244
  eagle **9**:1239, 1244,
    1250, 1251
  electric/torpedo
    9:*1238*
  electroreceptors **6**:*787*
  manta/devil **9**:*1244*,
    1249
  reproduction **9**:1251
  scales **9**:1283
  spotted eagle **9**:*1241*
  thornback **9**:*1250*
razorbill, wing bones
  7:874
Rebutia **8**:1059
receptor cells **6**:786
rectal glands **1**:130;
  9:*1234*, 1250
rectum
  insect **2**:183
  vertebrate **2**:182
rectus abdominalis
  7:936
red deer (wapiti)
  7:*964–983*
  bellowing **7**:*978*
  circulatory and
    respiratory systems
    7:965, *977–978*
  digestive and excretory
    systems **7**:965,
    979–980
  external anatomy
    7:965, 967–969

metabolic slow-down
  7:978
muscular system **7**:965,
  973–974
nervous system **7**:965,
  975–976
reproductive system
  7:965, *981–983*
skeletal system **7**:965,
  970–972
subspecies **7**:968
redwood **8**:*1138*
  California/coastal
    8:1139, 1142, 1144,
    1145, 1146
  dawn **8**:1140
reedbuck **10**:*1366*
reflexes **6**:796
regeneration **7**:988
  starfish and **9**:1228
regurgitation
  by eagles **2**:230
  by wolves **10**:1399,
    1402, 1403
  by woodpeckers
    10:1415, *1417*
reindeer *see* caribou
repletes **1**:33
reproductive system
  7:*984–1005*
  asexual reproduction
    7:985, *986–989*
    apomixis **7**:*988*, 989
    budding **7**:988
    parthenogenesis
      7:*986*, 988–*989*
    pedal laceration
      8:1094, *1095*
    polyembryony
      7:986
    regeneration **7**:988
    vegetative **7**:989
    see also fission
  sexual reproduction
    7:*985*, 990–993
    hermaphroditism
      7:*998*

reproductive
  structures in
  animals **7**:*994–999*
reproductive
  structures in plants
  7:*1000–1003*
  timing of **7**:997
  zygote
    development
    7:*1004–1005*
reptiles **2**:148–149;
  3:*396*; **5**:*624*;
  9:1195, *1266*, 1267
  brain **6**:*795*
  calcium regulation
    2:268
  excretion by **2**:186
  heart **1**:*110*, 112
  internal fertilization
    9:1279
  respiratory systems,
    ventilation
    8:1027–1028
  in salt water **2**:161
  skeletal system **9**:1160
    jaw **9**:1170
Reptilia **2**:*148*; **3**:*396*;
  5:*624*; **9**:*1266*
repugnatorial glands
  3:340
resilin **4**:543
resin **8**:1144–*1145*
respiration
  aerobic **4**:472; **5**:677,
    678; **7**:878; **8**:1012
  anaerobic **5**:678;
    8:1012
    by burrowing sea
      anemones **8**:1091
    by turtles **9**:1204
  buccopharyngeal
    6:*809*
  cellular **4**:566; **7**:910
  cutaneous (through
    the skin) **6**:804,
    809; **8**:1013–1014,
    1016

respiratory system
  8:*1012–1031*
  book lungs **8**:1015,
    1022, 1080;
    9:1257, *1261*
  breathing underwater
    8:1013, *1016–1018*
  breathing air using
    lungs **8**:1013,
    1019–1020
  gas exchange **8**:*1012*,
    1013–1015
  maximizing efficiency
    8:1013, *1029–1031*
  pigments **8**:1013, 1015,
    1023–1025
  see also hemoglobin;
    myoglobin
  tracheal system **8**:1013,
    1021–1022
  see also tracheae
  ventilation
    8:*1026–1028*
  see also respiration
rete mirabile **1**:120;
  3:*369*, 391;
  7:878
reticulum **2**:177; **3**:*370*;
  7:980; **10**:1375,
  1376
retina **4**:563
  of albatrosses **1**:16
  of chimpanzees **1**:99
  of eagles **2**:229
  of fruit bats **3**:304
  of humans **6**:*786, 788*
  of lions **5**:675, *676*
  of mammals **6**:*791*
  of ostriches **6**:840
  of sailfish **8**:1069
retractor muscles, of
  clams **3**:*350*–351
Retropinnidae **9**:1281
Retroviridae **10**:1314
retroviruses **10**:*1323*
rhabdomeres **9**:1260
Rhabdoviridae **10**:1314

Rhamnales **1**:*36*, 37
rhea **6**:*834*, 835
Rheiformes **6**:*834*
rheumatic fever,
    streptococcal **5**:622
rhinarium **6**:737; **7**:968
Rhinastus latesternus
    10:*1345*
Rhinobatiformes **9**:*1238*
rhinoceros **8**:**1032–1047**;
    10:*1418*, 1419
  black **8**:*1032*, 1033,
    1035–1036, *1037*,
    1039, *1041*, *1047*
  circulatory and
    respiratory systems
    8:1033, *1043*
  digestive and excretory
    systems **8**:1033,
    1044–1045
  external anatomy
    8:1033, *1034–1037*
  horns **8**:*1035–1036*
  Indian **8**:*1032*, 1035,
    1036, *1037*, 1047
  Javan **8**:*1032*, 1033,
    1034, 1035, *1037*
  muscular system
    8:1033, *1040–1041*
  nervous system **8**:1033,
    1042
  prehistoric **8**:*1036*
  reproductive system
    8:1033, *1046–1047*
  skeletal system **8**:1033,
    1038–1039;
    10:*1426*
    hind limb bones
      9:*1175*
  subspecies **8**:1037
  Sumatran **8**:*1032*,
    1033, 1034, 1035,
    1036, *1037*, 1046
  white **8**:*1032–1047*
Rhinoceros
  sondaicus **8**:*1032*
  unicornis **8**:*1032*

Rhinocerotidae **8**:*1032*;
    10:*1418*
Rhinodermatidae **1**:*54*
Rhizanthella gardneri
    6:827, *828*
rhizomes
  fern **2**:279–280, 282,
    285
  grass **6**:746–748, 751
  orchid **6**:826–827, *828*
  potato **7**:918, *919*,
    920–921, 989
  Venus flytrap **10**:1302,
    1307
rhizomorphs **6**:*776*
Rhizopoda **1**:21, 22;
    8:*1126*
Rhizostomae **5**:*636*, 638
Rhodophyceae **8**:*1126*
Rhodophyta **8**:*1126*,
    1127
rhomboideus muscle
    8:1040–1041;
    10:*1370*
rhopalia **5**:*643*; **6**:*789*
Rhyacotritonidae **6**:801
Rhyncophthirina **5**:*692*,
    693
ribonucleic acid *see*
    RNA
ribosomes **1**:*69*; **7**:984
  bacterial **1**:50, *70*
  rice **6**:752
Rickettsia **5**:607
rigor mortis **6**:759
rivers and lakes, common
    hippopotamuses
    4:510–511
RNA **1**:65, 80, 81
  and the first forms of
    life **7**:984
  messenger (mRNA)
    1:70, 80, 81–82, *83*;
    7:984; **10**:1320,
    1321
  ribosomal (rRNA)
    1:50, 82

  transfer (tRNA)
    1:*82–83*
Rodentia **4**:*476*; **6**:727;
    7:*898*, *946*; **9**:*1208*,
    1209
rodents **4**:*476*, 477;
    7:*898*, *946*, 947;
    9:*1208*, 1209
  ancestors **7**:949
  cavylike **7**:*898*, 899,
    907, *946*, 947, 952
  mouselike **7**:*898*, 907,
    946; **9**:*1208*
  squirrel-like **7**:*898*,
    907, *946*; **9**:*1208*
  tail **7**:955
  teeth **4**:482; **7**:905
Rodhocetus **2**:192
rods and cones **6**:*786*,
    791; **8**:1069;
    9:1274, 1287;
    10:1396
Romerolagus diazi **4**:*476*
roots **6**:*831*
  aerial **6**:*831*
  apple tree **1**:39
  buttercup **1**:*65*
  differ from stems **7**:920
  monocot **6**:826–827
  potato **7**:918–920,
    924–925
  as storage organs **6**:826
  Venus flytrap **10**:1302
rorquals **3**:375
  calves **3**:395
  feeding **3**:392–393
Rosa **1**:37
Rosaceae **1**:*36*, 37
Rosales **1**:*36*, 37
rose **1**:*36*, 37, 39
Rosidae **1**:*36*, 37;
    10:*1300*
rosids **1**:*36*, 37
rostral organ **1**:125, *128*
roundworm (nematode)
    6:774
Rousettus spp. **3**:*292*

royal jelly **4**:532
Rubivirus **10**:1314
rumen **2**:177; **3**:*370*,
    371; **7**:980;
    10:1375–1376,
    1377
Ruminantia **3**:*356*;
    4:*506*; **7**:964
ruminants **3**:*356*; **4**:*506*,
    522; **7**:964,
    979–980; **10**:*1360*
  sleep **10**:1373
  stomach **2**:177; **7**:979
runners **1**:43
rush **6**:*744*
Russia, sables **10**:1329
rust fungi **6**:771

S

sable **10**:1329
  horns **10**:*1365*
Saccharum officinarum
    6:744
Saccopharyngidae **3**:*436*
Saccopharyngiformes
    3:*436*, 437
Saccopharyngoidei **3**:*436*
Saccopharynx **3**:438–439,
    441, 442, 444
sacrum **4**:558; **6**:732;
    7:935; **9**:1173;
    10:1391
Sagartia troglodytes **8**:1094
sagittal crest **1**:92; **4**:514
Sagittariidae **2**:*220*
saguaro cactus
    8:**1048–1059**
  birds and **8**:*1057*
  "boots" **8**:*1057*
  external anatomy
    8:*1051*
    flowers **8**:*1051*,
      1058–1059
    fruit **8**:*1052*, 1059
  internal anatomy
    8:*1055*, 1056
  pollination **8**:*1058*

roots **8**:1054
seed dispersal **8**:1058
saiga **10**:*1360*
sailfish **8**:**1060–1073**
  Atlantic **8**:*1060, 1066*
  circulatory and
    respiratory systems
    8:1061, *1070–1071*
  digestive and excretory
    systems **8**:1061,
    1072
  external anatomy
    8:1061, *1062–1064*
  Indo-Pacific **8**:*1060*
  muscular system
    8:1061, *1066–1067*
  nervous system **8**:1061,
    1068–1069
  reproductive system
    8:1061, *1073*
  skeletal system **8**:1061,
    1065
  swim bladder **8**:1072
salamander **1**:*54*; **6**:*800*
  alpine **6**:802
  Asian land **6**:801
  Chinese giant **6**:801
  circulatory and
    respiratory systems
    6:*808*–809
  European **6**:*800*, 801
  fire **6**:802, 804, 807
  Japanese giant **6**:801
  larvae **6**:*804*
  lungless **6**:*800*, 801,
    806, 808
  mole **6**:*800*, 801, *803,*
    804
  movement **6**:805, 806
  neoteny **6**:811
  nervous system **6**:*781*
  northern red **6**:*801*
  Pacific giant **6**:801
  senses **6**:807
  skeleton **6**:805, 806;
    9:1159
  spotted **6**:*803*

tiger **6**:801, *806*, 811
torrent **6**:801
Salamandridae **6**:*800,*
  801, 802
Salangidae **9**:1281
saliva **2**:173, 274
salivary glands **2**:274
  of anteaters **3**:325–326
  of centipedes **3**:336
  of humans **4**:570
  of lice **5**:697, *698*
salmon **8**:1066, 1067;
  9:*1280*, 1281
  Atlantic **9**:*1280*, 1281
  Pacific **9**:1281, 1293
Salmonella typhosa **5**:607
Salmonidae **9**:*1280*, 1281
Salmoniformes **9**:*1280,*
  1281
Salmo salar **9**:*1280*
salt glands
  albatross **1**:*18*
  crocodile **2**:*152*, 161
  marsh grass **6**:748
  penguin **7**:879
Salticidae **9**:1260
Salvelinus fontinalis
  9:*1280*
Samotherium boissieri
  3:361
sand dollar **9**:*1224,*
  1225
sandworm, jaws **2**:178
saprotrophs, fungi as
  6:774
sapsucker **10**:1415
  yellow-bellied **10**:1406,
    1415
Sarcodina **1**:*20*, 21;
  2:*164*
sarcodines **2**:*164*
sarcolemma **6**:757
sarcomeres **6**:756, 757,
  759
Sarcophagidae **4**:*538*
sarcoplasmic reticulum
  6:757, 760

sarcopterygians
  9:1158–1159
Sarcopterygii **1**:*122*, 123;
  3:*436*; **8**:*1060*;
  9:*1280*
sargassum **8**:1126, 1134
Sarracenia **10**:*1300*
Sarraceniaceae **10**:*1300,*
  1301
Saturnidae **4**:*494*
Sauria **5**:*624*
Saurosuchus imperator
  2:154
savanna, giraffes **3**:366
sawfish **9**:*1238*
sawfly **1**:*26*, 27; **4**:*526*
scales
  cosmoid **9**:1283
  ctenoid **9**:*1283*
  cycloid **9**:*1283*
  ganoid **9**:1283
  placoid (denticles)
    4:*463*; **9**:*1283*
scallop **3**:*342*, 345, 353
  escape response **9**:1234
  eyes **3**:*353*
  jet propulsion **3**:*351*
Scandentia **6**:727
scape **1**:31
Scaphopoda **3**:*342*;
  6:812
scapulae (shoulder
    blades) **9**:1174
  of chimpanzees **1**:93
  of giant anteaters **3**:321
  of giraffes **3**:363
  of montremes **7**:889
  of zebras **10**:1426
scapulohumerus muscles
  1:15
scarab **10**:*1344*
scats **3**:426
scent
  cat **7**:939
  deer **7**:969, *976*
  giant anteater **3**:327
  hyena **5**:595, 600–*601*

manatee **5**:716
moose **7**:*967*
otter **6**:854, 860
porcupine **7**:909
rat **7**:956, 963
weasel **10**:1335
wildebeest **10**:*1363*
wolf **10**:*1399*
schistosomiasis **7**:987
Schizophora **4**:*538*, 539
Schwann cells **6**:*785*
scions **1**:45
Sciuridae **9**:*1208*, 1209
Sciurinae **9**:*1208*, 1209
Sciurognathi **4**:547;
  7:*898, 946*; **9**:*1208*,
  1209
Sciurus **9**:*1208*, 1209
Scleractinia **8**:*1084*
sclerenchyma cells **1**:67
Sclerocactus papyracanthus
  8:1053
Scolioidea **4**:*526*
Scolopendramorpha
  3:*332*
Scombroidei **8**:*1060,*
  1061
scombroids **8**:1061
scopulae **9**:1255–1256
scorpion **8**:**1074–1083**;
  9:*1252*
  ancestors **8**:1077
  black **8**:*1075*
  circulatory and
    respiratory systems
    8:*1022*, 1075, *1080*
  digestive and excretory
    systems **8**:1075,
    1081
  emperor **8**:*1076, 1077,*
    1079–1083
  external anatomy
    8:1075, *1076–1077*
  muscular system
    8:1075, *1078*
  nervous system **8**:1075,
    1079

scorpion (continued)
  reproductive system
    8:1075, 1082–1083
  sting and venom
    8:1076, 1081
  see also wind scorpion
Scorpiones 8:1074;
  9:1252
scorpionfly 4:538
Scrophulariales 7:916
scrotum 7:994
Scutigeromorpha 3:332
Scyliorhinidae 4:460
scyphistoma 5:637, 638,
  646, 647; 7:986,
  987
Scyphozoa 5:636;
  8:1084
sea anemone
  8:1084–1095
  burrowing 8:1091,
    1094
  common plumose
    8:1085, 1086,
    1087, 1088, 1095;
    9:1166
  crabs' use of 1:135;
    8:1087
  digestive and excretory
    systems 8:1085,
    1092–1093
  diploblastic layers
    8:1089
  external anatomy
    8:1085, 1086–1087
  giant 8:1086
  internal anatomy
    8:1085,
    1088–1089; 9:1166
  lacking blood 3:354
  larvae 8:1094
  muscles 6:755
  nervous system 8:1085,
    1090
  northeast Pacific
    8:1095
  plumose 8:1085

reproductive system
  8:1085, 1094–1095
respiratory system
  8:1085, 1091
tentacles 8:1086, 1087
  defense 8:1090
seabed
  angel sharks 4:463
  hagfish 4:547
sea cat 6:848
sea cow 2:244, 245
  blubber 2:192
  Steller's 5:700
see also dugong; manatee
sea cucumber 9:1224,
  1225, 1228
sea daisy 9:1224, 1225
sea dragon 8:1100
sea fan 8:1085, 1087
seahorse 8:1096–1109
  camouflage 8:1100
  circulatory and
    respiratory systems
    8:1097, 1104–1105
  clicking noises 8:1103
  digestive and excretory
    systems 8:1097,
    1106–1107
  evolution 8:1100
  external anatomy
    8:1097, 1099–1100
  eyes and vision 8:1100
  gills 8:1100, 1105
  golden 8:1102
  long-snout 8:1107
  nervous system
    8:1097
  number of species
    8:1098
  as predators and prey
    8:1107
  pygmy 8:1098, 1107
  reproductive system
    8:1097, 1108–1109
  skeletal system 8:1097,
    1101–1102
  snout of 8:1107

spotted 8:1097, 1099,
  1101, 1103, 1104,
  1105, 1106, 1109
thorny 8:1098
tiger tail 8:1105
White's 8:1102
seal 3:410; 6:844;
  8:1110–1125;
  10:1382
blubber 2:192
circulatory and
  respiratory systems
  8:1111, 1120–1121
common 8:1115
crabeater 8:1122
digestive and excretory
  systems 8:1111,
  1122–1123
diving by 8:1013,
  1117, 1121
eared 8:1111, 1113,
  1115, 1116
see also sea lion
elephant 8:1013, 1113,
  1114, 1115, 1118,
  1121, 1122, 1123,
  1125
external anatomy
  8:1111, 1112–1115
eyes and vision 8:1114,
  1118, 1119
fasting by 8:1123
fur 8:1110, 1111,
  1115, 1116
gray 8:1123, 1125
harbor 8:1111, 1112,
  1113, 1114–1115,
  1116, 1117, 1120,
  1122, 1124
harp 8:1114
Hawaiian monk 8:1112
hearing underwater
  8:1119
hooded 8:1121
leopard 8:1115
muscular system
  8:1111, 1117

nervous system 8:1111,
  1118–1119
northern 8:1110, 1111
reproductive system
  8:1111, 1124–1125
ringed 8:1112, 1113,
  1116, 1117, 1122
skeletal system 8:1111
sounds 8:1121
southern 8:1110, 1111
tasting saltiness of
  seawater 8:1119
true (earless) 8:1110,
  1113, 1115, 1116,
  1117
Weddell 8:1112, 1113,
  1117, 1120, 1121
whiskers 8:1114, 1119
white-coated true
  8:1111
sea lettuce 8:1126, 1129,
  1135
  reproduction
    8:1136–1137
sea lily 9:1225, 1228
sea lion 3:410; 8:1110,
  1111; 10:1382
  blubber 2:192
  leg muscles 8:1117
sea pen 8:1087
sea slug, and anemone
    stinging cells
    8:1092
sea spider 9:1252
sea squirt, pharynx
    2:176
sea star, sunflower
    9:1225, 1234
see also starfish
sea urchin 9:1224, 1225,
  1228
  growth 9:1167
  sea otters and 6:859
  skeletal system 9:1164,
    1167
  slate pencil urchin
    9:1228

sea wasp **5**:638
seaweed **8:1126–1137**
  caenocytic **8**:1133
  external anatomy
    8:1127, *1128–1131*
  fossil **8**:1130
  internal anatomy
    8:1127, *1132–1133*
  niches **8**:1131
  reproductive system
    8:1127, *1134–1137*
sebaceous glands, puma
    7:933
secretary bird **2**:*220*
secretin **2**:272
sedge **6**:*744*
seeds, plant **2**:276–277;
    7:1003
  dispersal **6**:740, 753;
    7:1003; **8**:1058,
    1148–1149
  grass **6**:752–*753*
  orchid **6**:778
  useful **6**:752
Selachii **9**:*1238*
self-tolerance **5**:620
Semaeostomae **5**:*636*,
    638
semen **7**:995
semicircular canals
    9:1288
semi-digitigrade posture
    4:513
seminal receptacle **3**:341;
    9:*1264*
seminiferous tubules
    7:*994*
semispinalis capitis
    10:1412
semitendinosus muscles
    8:1117
sensilla **4**:497
sensory adaptation
    6:786–787
sensory organs **6**:787
  loss during evolution
    6:789

sensory transduction
    6:787
sepals
  orchid **6**:*832*
  potato flower **7**:927
Sepioida **6**:*812*, 813
Sepiolida **6**:813
sequoia **8:1138–1149**
  external anatomy
    8:1139, 1140–*1143*
  internal anatomy
    8:1139, *1144–1145*
  reproductive system
    8:1139, 1146–*1149*
Sequoiadendron giganteum
    8:*1138*, 1139
serous glands **2**:274
Serpentes **3**:*396*; **5**:*624*
serratus anterior muscles
    10:1412
serratus ventralis muscles
    10:1370
Sertoli cells **7**:*994*, 998
sesamoids, of giraffes
    3:363
setae **1**:137
  of earthworms **2**:235,
    236–237, 239, 243
severe combined
    immunodeficiency
    (SCID) **5**:*622*, 623
sex changing
  bivalves **3**:355
  fish **4**:459; **7**:*998*
  determination
    7:991–992
  temperature and
    9:1206–1207
sexual selection **10**:1379
Seychelles, giant tortoises
    9:*1267*
S-gene **1**:45
shark **4**:461; **9**:*1238*,
    1239
  angel **4**:463; **9**:1242
    spiracles **4**:*473*
  buoyancy **9**:1249

cat **4**:463
  electroreceptors **6**:*787*
  frilled **4**:463
  great white **4**:*467*, 469,
    472; **8**:1027
  ground **4**:*460*, 461
  hammerhead *see*
    hammerhead shark
  requiem **4**:*460*
  respiratory system,
    ventilating gills
    8:*1027*
  saw **9**:1242
  scales **9**:1283
  skeletal system **4**:461,
    464–465; **9**:1169
    vertebrae **9**:1172
  weasel **4**:460
shearwater **1**:*10*
sheep **3**:357
sheep strike (myiasis)
    4:548
shell, giant clam
    3:*344*–345, 349
shell gland **7**:881
shipworm **3**:*342*
shock, anaphylactic **5**:616
shoulder girdle, of turtles
    9:1200
shrew, tree *see* tree shrew
shrimp **1**:133; **5**:*684*
Sigmodontinae **7**:*946*,
    947
silk, spider **9**:1264
Simplexvirus **10**:1314
singing, whales **3**:386,
    394
sinoatrial (SA) node
    1:116; **6**:767
sinus venosus **8**:1070,
    1104, 1105; **9**:1248
Siphonaptera **4**:*538*
siphonoglyphs **8**:1088
siphonostele **2**:282
siphuncle **6**:813, 815
siren **6**:800, 803
Sirenia **2**:*244*; **5**:*700*

sirenians **5**:700–701
see also dugong; manatee
sitatunga, hooves
    10:*1366*
Sivatherium **3**:361
skate **9**:*1238*, 1239
  electroreceptors
    9:*1247*
  reproduction **9**:1251
skeletal system
    9:1156–1175
  evolution **9**:1157–1160
  hydrostatic/hydraulic
    3:323; **9**:*1162*–1164
  of caterpillars **4**:501
  of earthworms
    2:239; **9**:*1163*–1164
  of jellyfish **5**:642;
    9:*1162*
  of sea anemones
    6:*755*
  invertebrate
    9:*1162*–1167
  plant **9**:*1161*
  vertebrate
    9:*1168–1175*
    skull **9**:*1170–1171*
  see also bone; cartilage
  see also exoskeleton
skimmer **2**:*208*
skin **5**:609
  human **4**:*553*
  lizard **5**:*631*
skipper **4**:*494*
Skrjabingylus nasicola
    10:1332
skull, vertebrate
    9:*1170–1171*
skunk, odor **2**:275
sleep, ruminants and
    10:1373
sleeping sickness **1**:21
slime mold **6**:770;
    8:*1126*
sloth **3**:*312*, 313;
    9:1176–1193
  ancestors **9**:*1182*

sloth *(continued)*
  brown-throated three-
    toed **9**:*1176*, 1180,
    1183, *1189*, *1191*
  circulatory and
    respiratory systems
    9:1177, *1188–1189*
  claws **3**:316
  digestive and excretory
    systems **9**:1177,
    1190–1191
  external anatomy
    9:1177, *1178–1183*
  fur
    cyanobacteria in
      9:*1178*, 1182, 1183
    moths in **9**:1182
  giant **3**:320
  Hoffman's two-toed
    9:*1176*, 1179–1180,
    1183, 1193
  maned three-toed
    9:*1176*, 1180,
    1182–1183, 1190
  muscular system
    9:1177, *1186*
  nervous system **9**:1177,
    1187
  pale-throated three-
    toed **9**:*1176*, *1177*,
    1178, 1180, 1183,
    1184, *1186*, *1188*,
    1190
  reproductive system
    9:1177, *1192–1193*
  skeletal system **9**:1177,
    1184–1185
  southern/Linne's two-
    toed **9**:*1176*,
    1179–1180, *1187*
  three-toed **3**:*312*, 313;
    9:*1176*, 1177,
    1179, *1180*, 1181,
    1185
  two-toed **3**:*312*, 313;
    9:*1176*, 1177,
    1179, *1180*, 1185
slug **3**:342
smell, sense of **6**:787,
  790
smut fungi **6**:771
snail **3**:342
  digestive system **2**:179
  lung **8**:1019
snake **3**:*396*, 397; **5**:*624*;
    9:*1194*
  blind **3**:*403*
  file *see* file snake
  pipe **3**:*403*
  sex determination
    9:1207
  teeth **2**:174
  thread **3**:*403*
  venom **2**:275
  vision **9**:1274
snakefly **4**:*538*
snake mackerel **8**:*1060*
snapping turtle
    9:**1194–1207**, *1266*
  alligator **9**:*1195*,
    1197–1200, *1205*,
    1206, 1207
  circulatory and
    respiratory systems
    9:1195, *1203–1204*
  common
    9:*1196–1197*, 1206,
    1207
  digestive and excretory
    systems **9**:1195,
    1205
  external anatomy
    9:1195, *1196–1198*
  muscular system
    9:1195, *1201*
  nervous system **9**:1195,
    1202
  reproductive system
    9:1195, *1206–1207*
  skeletal system **9**:1195,
    1199–1200
snout, dental **6**:733
Solanaceae **7**:*916*, 925
Solandra **7**:925
Solanum **7**:917
  lycopersicum **7**:*916*
  tuberosum **7**:*916*
Solenidae **3**:*342*
soleus muscle **4**:561
Solpugida **8**:*1074*, 1075
solpugids **8**:1075
somatostatin **2**:267, 270
somatotropin **2**:266, 270
sori, fern **2**:284, *285*
sound production
  baleen whales **3**:388
see also infrasound
South America
  anacondas **3**:397
  armadillos **9**:*1181*
  hummingbirds **5**:580
  manatees **5**:700
  maned wolves **10**:*1385*
  piculets **10**:1405
  porcupines **7**:900, 902,
    903, 912, *913*
  tinamous **6**:835
South Carolina, Venus
    flytrap **10**:1301
Spain, weasels **10**:1340
sparrow, Lincoln's **7**:*999*
Spartina **6**:*744*, 745, 750
  alterniflora **6**:*744*, 750
  anglica **6**:*744*, 750, *751*
  maritima **6**:*744*, 750
spearfish **8**:*1060*, 1064
  Mediterranean **8**:1064
speciation, centipedes
    3:335
spectacle scales **3**:399
sperm (spermatozoa)
    1:67, 87; **7**:*990*,
    993, *994*, 995–996
  competition **3**:395
spermaceti organ **2**:203
spermathecae **4**:504
  of ants **1**:34
  of earthworms **2**:237
  of houseflies **4**:549
  of newts **6**:811
  of weevils **10**:1357
spermatia **8**:1136
spermatids **7**:993
spermatocytes **7**:993
spermatogenesis **7**:*994*
spermatogonia
    7:992–993, 994
spermatophores **3**:336,
    341; **4**:504; **5**:691,
    699; **6**:*810*; **8**:*1082*
spermatozoids **8**:1137
Spermophilus **9**:*1208*
Spheniscidae **7**:*868*
Sphenisciformes **7**:*868*
Spheniscus **7**:*868*
Sphenophyta **2**:*276*
sphincter muscles **6**:763,
    764
Sphingidae **4**:*494*, 495
Sphingoidea **4**:495
Sphyraenoidei **8**:*1060*
Sphyrnidae **4**:*460*
spider **8**:*1074*; **9**:1253
  crab **9**:1256
  digestive system **2**:*179*
  feeding **2**:175
  funnel-web **9**:1263
  giant trapdoor **9**:*1252*
  jumping **9**:1260
  mechanoreceptors
    6:*791*; **9**:*1259–1260*
  mouthparts **2**:175
  respiratory system
    8:*1022*
  water **8**:*1018*
  world's largest **9**:*1253*
see also tarantula
spike moss **2**:285
spinal cord, human **6**:*796*
spinnerets **4**:499; **9**:1254,
    1256, 1257, 1264
Spinulosida **9**:*1224*
spiracles **8**:1021
  of dragonflies **2**:213,
    217
  of fish
    bottom-living sharks
    4:*473*

rays **9**:1240, *1248*
of lice **5**:*696*
of moths **4**:503
of myriapods and
    centipedes **3**:333,
    336, 338
of scorpions **8**:1078
of weevils **10**:1353,
    1355–1356
Spiranthoideae **6**:*824*,
    825
Spirochaetes **1**:47
Spirulida **6**:813
sponge
    skeletal system **9**:1162
    ventilation **8**:*1026*
sporangia
    fern **2**:*283*, 284, *285*
    fungal **6**:779
sporangiospores **6**:777
spores
    bacterial **1**:*53*
    fern, dispersal **2**:284
    fungal **6**:775, 777,
        778–779
    seaweed **8**:1134, 1135
sporocytes **2**:*283*, 284
sporophyte **7**:993, 1000,
    1002
    fern **2**:277; **7**:*993*
    seaweed **8**:*1134*
sporozoans
    (apicomplexans)
    1:21, 24
springtail **4**:494–495
spyhopping, whales
    3:386
Squamata **3**:*396*; **5**:*624*;
    9:*1194, 1266*
squamates **3**:396; **5**:*624*;
    9:*1266*
squamosal bone, in
    chameleons **5**:628,
    629
squid **3**:*345*; **6**:*812*
    circulatory system
        1:107

color change **6**:816
dumpling **6**:813
ink **6**:817
nervous system
    eyes **6**:788
    giant nerve cells
        6:783, *820*
    vampire **6**:*812*
squirrel **9**:**1208–1223**
    Abert's **9**:1212
    African pygmy **9**:1210
    arctic ground **9**:1211,
        1215
    Arizona gray **9**:*1221*
    black giant **9**:1210,
        1211
    circulatory and
        respiratory systems
        9:1209, *1219*
    digestive and excretory
        systems **9**:1209,
        1220–1221
    Douglas (chickaree)
        8:*1148*
    eastern gray **9**:1210,
        1220, 1223
    Eurasian red
        9:*1210–1216*,
        *1218, 1219, 1220,
        1222*, 1223
    evolution and
        relationships **9**:1215
    external anatomy
        9:1209, *1210–1213*
    feet **7**:950; **9**:1211
    feeding on fungus
        9:1221
    flying **9**:*1208, 1213*
        tail **7**:955; **9**:*1217*
    giant flying **9**:*1212,
        1213*
    gray **9**:*1212, 1223*
    groove-toothed **9**:1212
    ground **9**:*1208*,
        1211–*1212*, 1220
    muscular system
        9:1209, *1216–1217*

nervous system **9**:1209
northern flying **9**:1223
pygmy tree **9**:1217
red **9**:*1209*, 1221
reproductive system
    9:1209, *1222–1223*
rock **9**:1223
scaly-tailed **9**:*1208*,
    1209, 1213
shrew-faced ground
    9:1212
skeletal system **9**:1209,
    1214–1215
southern flying **9**:1213,
    1217
tail of **9**:*1212*, 1217
tree **9**:*1208*,
    1211–*1212*, 1220
Sri Lanka, chameleons
    5:624
stable fly **4**:539
stamens **1**:43; **6**:751–752;
    7:*1001*
stannius gland **2**:268
stapes **4**:555; **9**:1170
Staphylinidae
    10:1347–1348
staphylococci **5**:608
starfish **9**:**1224–1237**
    circulatory and
        respiratory systems
        9:1225, *1233*
    crown-of-thorns
        9:*1224*, 1235
    digestive and excretory
        systems **9**:1225,
        1234–1235
    external anatomy
        9:1225, *1226–1228*
    internal anatomy
        9:1225, *1229–1231*
    nervous system **9**:*1232*
    reproductive system
        9:1225, *1236–1237*
    skeletal system **9**:1164
    sunflower **9**:*1224*,
        1226

see also sea star
statocysts **1**:137; **6**:820
statoliths **5**:643
Stauromedusae **5**:*636*
stem cells **1**:108; **7**:*959*
stems
    differ from roots
        7:918
    orchid **6**:827, 829
    potato **7**:918, *919*, 923
    underground *see*
        rhizomes
Stenocereus
    eruca **8**:1050
    thurberi **8**:1050
stercoral pocket **9**:1257
stereocilia **6**:*789*
stereom **9**:1225
sternum
    of birds **9**:1173
        albatross **1**:14
        eagle **2**:226
        hummingbird
            5:*585*
        ostrich **6**:*838*, 839
        woodpeckers
            10:1410
    human **4**:558
    of spiders **9**:1254
steroids **2**:265, *271*
stickleback **8**:1100
stigmata **1**:*43*; **7**:*927*
stingray **9**:**1238–1251**
    blue-nose **9**:1244
    blue-spotted **9**:1244
    blunt-nose **9**:1244
    circulatory and
        respiratory systems
        9:1239, *1248*
    diamond **9**:1250, 1251
    digestive and excretory
        systems **9**:1239,
        1249–1250
    external anatomy
        9:1239, *1240–1241*
    nervous system **9**:1239,
        1245–1247

stingray *(continued)*
  reproductive system
    9:1239, *1251*
  round **9**:1247
  skeletal and muscular
    systems **9**:1239,
    1242–1244
  southern **9**:*1239, 1240,*
    1241, 1244, 1251
  venomous sting **9**:*1241*
  yellow **9**:*1250*
stinkhorn **6**:771, 779
stipe **8**:1128, *1129*–1130
stoat *see* ermine
stolons **6**:748, 751; **7**:989
stomach **2**:177–178
  of ants **1**:32
  cardiac **1**:139; **5**:689;
    9:1234, *1235*
  of carnivores **3**:325
  of cows **5**:*659*
  of crabs **1**:139
  of deer **7**:*979*, 980
  of eagles **2**:230
  of kangaroos
    5:*659*–660
  of lobsters **5**:689
  pyloric **1**:139; **5**:689;
    9:1234
  of ruminants **2**:177;
    7:979
  of tarantulas **9**:*1258*,
    1262
stomata **1**:42; **6**:748, 749,
    829; **8**:*1145*
  cactus **8**:1054,
    1056–1057
  fern **2**:281–282
  Venus flytrap **10**:1306
stone canal **9**:1229–1230
stonefly, nymphs **8**:1017
stork **5**:695
storm petrel **1**:*10*, 11
stramenopiles **2**:*164*
strawberry **1**:37, 38, 43,
    44; **7**:989
Strepsiptera **4**:*538*

strepsipterans **4**:542
  halteres **2**:211
Strepsirrhini **4**:*550*
Streptococcus pneumoniae
    5:*607*
Strigamia maritima **3**:335
strobilae **7**:986–987
stromatolites **8**:1130
Struthio camelus **6**:*834*
Struthioformes **6**:*834*
Stylasterias forreri **9**:1227
style (animal) **2**:179
  crystalline **3**:352
styles (plant) **1**:*43*; **6**:752
stylets, of lice **5**:697, *698*
subesophageal ganglion
    2:215
sublingual glands **2**:274
suborbital glands **7**:976
sugarcane **6**:744
Suidae **4**:*506*
Suiformes **4**:*506*; **7**:*964*;
    10:*1360*
Suinae **4**:*506*
sulci **4**:565
Sumatra, porcupines
    7:900
sundew **10**:*1300*, 1301,
    1303–1304, *1312*
sunflower, seeds **6**:752
sun scorpion *see* wind
    scorpion
sun star, common **9**:*1224*
superficial masseter
    muscle **9**:1216
superior rectus muscle
    8:1069
supracoracoideus
    (pectoralis minor)
    muscles **6**:*839*;
    7:875; **10**:1412
supracoracoid muscles
    5:586
supraorbital processes
    4:482
swallower **3**:*436*
sweat glands **2**:274–275

swift **5**:*580*
swim bladder **6**:765
  in coelacanths **1**:129
  in sailfish **8**:1072
  in sea horses **8**:*1104*
  in trout **9**:1286
swimmerets **5**:687, 691
swimming
  dolphins **2**:192, 198
  fish **9**:1285
    billfish
      8:1066–1067
    coelacanths **1**:*127*
    guitarfish **9**:1243
    sharks **4**:464
    skates and rays
      9:*1243–1244*
    tuna **6**:764–765;
      8:1066–*1067*
  manatees **5**:708
  penguins **7**:*870–871*,
    873–874
  underwater **3**:379;
    6:848
  whales **2**:198; **3**:376,
    379, 384
swordfish **8**:*1060*, 1064
Syconycteris spp. **3**:*292*
Sylvilagus **4**:*476*
symbiosis
  ants "farming" other
    insects **1**:33
  and the evolution of
    eukaryotes **2**:170
  light-producing
    bacteria in fish
    3:439
  microorganisms in guts
    of lice **5**:697–698
  of sloths **9**:1191
  zooxanthellae in
    cnidarians **8**:1092
symmetry
  bilateral **4**:476; **10**:1300
  radial **5**:637; **7**:927;
    8:1085; **10**:1300
  pentaradial **9**:*1225*

Symphata **4**:*526*
Symphyla **3**:*332*
symphylans **3**:*332, 341*
Symphyta **1**:*26*
synapses **4**:562;
    6:784–785, *791*
syndactyl animals **5**:655
Syngnathidae **8**:*1096*
Syngnathiformes **8**:*1096*,
    1098
Syngnathinae **8**:*1096*,
    1098
Syngnathus **8**:1109
synsacrum **7**:873; **9**:1173;
    10:1411
syrinx **10**:1414
Syrphidae **4**:*538*
systemic lupus
    erythematosus
    5:623
systole **1**:117

T

T3 (thyroid hormone)
    2:267, 271
T4 (thyroid hormone)
    2:267, 271
Tachinidae **4**:*538*
Tachyglossidae **7**:*882*
Tachyglossus aculeatus
    7:*882*
tadpoles **1**:57, 61, 62, 63
  metamorphosis **1**:57,
    62, 63; **2**:267
  newt **6**:802–803
taenidia **8**:1021–1022
tails, prehensile, of
    monkeys **6**:727, 735
tamandua **3**:*312*, 313
  anal secretions **3**:327
  claws **3**:316
  collared **3**:316
  northern **3**:316, *317*
  skull **3**:*319*
  southern **3**:316, *328*
  tail **3**:317
  teeth **3**:321

Tamandua **3**:*312*
Tamias **9**:*1208*
Tanzania, chameleons
    5:624
tapetum lucidum **1**:124,
    128; **5**:676, 710;
    6:855; **8**:1118;
    9:1245
tapeworm **2**:*182*, 183
tapir **8**:*1032*; **10**:*1418*,
    1419
  skull **10**:*1426*
tapirfish **3**:*436*
Tapiridae **8**:*1032*;
    10:*1418*
tarantula **8**:1022;
    9:**1252–1265**
  circulatory and
    respiratory systems
    9:1253, *1261*
  digestive and excretory
    systems **9**:1253,
    1262–1263
  external anatomy
    9:1253, *1254–1256*
  goliath **9**:*1253, 1261*
  internal anatomy
    9:1253, *1257*
  king baboon **9**:*1263*
  Mexican red-kneed
    9:1260
  nervous system **9**:1253,
    1259–1260
  reproductive system
    9:1253, *1264–1265*
  skeletal and muscular
    systems **9**:1253,
    1258
tarpon **3**:*436*
tarsal organs **9**:1260
tarsier **4**:*550*; **6**:731
Tarsii **4**:*550*
tarsometatarsus **6**:839
tarsus **10**:*1425*
Taurotragus **10**:*1364*
Taxaceae **8**:1139
Taxodiaceae **8**:*1138*

tayra **10**:*1324*, 1339
T-bacteriophage virus
    10:*1320, 1321*
T-cell receptors (TCRs)
    5:608, 617, 620
T-cells *see* lymphocytes
tears **5**:611
teeth **2**:173–175
  canine **2**:174;
    3:418–419; **4**:*515*;
    5:678; **6**:*851*
  of chimpanzees
    1:*96*, 101, *103*
  of deer **7**:*969*, 971
  of horses **10**:1435
  of mandrills **6**:*732*,
    739, 740, 741
  of wolves **10**:*1391*,
    1398, *1399*
  carnassial **3**:419; **5**:596,
    665; **6**:*847, 851*;
    7:942; **10**:1332,
    1398
  diphyodont and
    polyphyodont
    5:713
  incisor **2**:174; **5**:*655*;
    6:741, *851*; **7**:905,
    950, 971; **8**:1044;
    9:*1215*; **10**:1398,
    1399
  lingual **4**:*450*
  mammal **2**:*173*, 174
    development **6**:*740*
  maxillary **9**:*1291*
  molars **2**:*173*, 174;
    10:1435
  pharyngeal **2**:178
  premolars **2**:174; **5**:596;
    8:1044; **10**:1435
  vomerine **9**:*1291*
  see also tusks
temperature control
  by albatrosses **1**:13
  by bats **3**:296
  by bears **3**:425
  by eagles **2**:231

  by hares and rabbits
    4:*479*
  by hippopotamuses
    4:521
  by honeybees **4**:535
  by kangaroos **5**:*658*
  by ostriches **6**:841
  by otters **6**:856
  by three-toed sloths
    3:313
  by tortoises **9**:1276
  by whales **3**:391
  see also countercurrent
    heat exchangers
temporal (temporalis)
    muscles **9**:1217;
    10:*1334*
tendons **3**:420; **6**:763;
    7:954
  in a kangaroo's hind
    legs **5**:656
tendrils, plant **9**:1161
Teredinidae **3**:*342*
tergites **1**:134; **6**:762
termite **1**:*26*, 29
  king and queen **1**:35
Termitomyces titanicus
    6:775
terrapin
  diamond-backed
    9:1207
  painted **9**:1204
testes
  human **4**:*572*, 573;
    7:*994*
  internal and external
    3:330; **8**:1047
testes-determining factor
    (TDF) **7**:992
testosterone **4**:571;
    7:992, 995, 998
  and birdsong **7**:999
  in hyenas **5**:601, 605
  in mandrills **6**:*730*, 743
Testudinata **9**:*1194*
Testudinidae **9**:*1194*,
    1266

Testudo **9**:1273
tetanus (lockjaw) **5**:*607*
tetraploid plangts **7**:927
tetraspores **8**:*1134, 1135*
Teuthoida **6**:*812*, 813
thalamus **4**:565;
    6:797–798
thallus **8**:1129
Thaumastochelidae
    5:*684*
Theales **10**:*1300*, 1301
theca (frustule), of
    diatoms **2**:166, *167*
theca cells **7**:995
Thelymitra spiralis **6**:828
Theraphosidae **9**:*1252*,
    1253
Thermodesulfobacteria
    1:47
thermogenesis,
    nonshivering **7**:875
thermogenic organs
    8:1069
thermoreceptors,
    tarantula **9**:1259,
    1260
thermoregulation *see*
    temperature control
thigmotropism **9**:1161
thirst **6**:787
thromboplastin **1**:119
thrombosis **1**:*117*
thumbs, opposable **1**:*97*;
    4:553; **6**:729, *733*
thylakoids **6**:*750*
thymocytes **5**:619
Thymollus arcticus **9**:*1280*
thymus **5**:619, *621*
thyroid gland **2**:267
thyroid hormone **2**:267
thyroiditis **5**:622
thyroid-releasing
    hormone (TRH)
    2:267
thyroid-stimulating
    hormone (TSH)
    2:267, 270–271

thyroxine **2**:267
tibia **9**:*1174*
tibiofibula **1**:59
Tiburonia granrojo **5**:638
tick **5**:698; **8**:*1074*–1075
Tiedemann's bodies **9**:1230
tiger **5**:*664, 668*
tillers **6**:748, 751
Tinamiformes **6**:*834*
tinamou **6**:*834*, 835
toad **1**:*54*, 55; **6**:*800*
  cane, poison glands **2**:*275*
  European common **1**:62
  Suriname **1**:63
  true **1**:*54*
toadfish, oyster **6**:765
toadstools **6**:*771*
tobacco **7**:*916*
tobacco mosaic virus **10**:*1316*, 1317
Togaviridae **10**:1314
toll-like receptors (TLRs) **5**:610–611, 619
tomato **7**:*916*
Tömösvary organ **3**:333
tongue
  of chameleons **5**:*630, 633*
  of giant anteaters **3**:322–*323*, 325
  of giraffes **3**:*359*, 364–365
tool use
  chimpanzees and **1**:*97*, 104
  elephants and **2**:257
Torpediniformes **9**:*1238*
torpor, hummingbirds **5**:588
tortoise **9**:*1194*, 1195, **1266–1279**
  Aldabra giant **9**:*1267*, 1277, 1279

circulatory and respiratory systems **9**:1267, *1275–1276*
desert **9**:*1271*
digestive and excretory systems **9**:1267, *1277*
external anatomy **9**:1267, *1268–1271*
Galápagos giant **9**:*1269, 1271–1279*
  shell shapes **9**:*1270*
hinge-back **9**:1268, 1273
Horsfield's/Russian **9**:*1268*
land **9**:1267, 1273, 1277
Madagascar plowshare **9**:*1279*
Mediterranean **9**:1273
muscular system **9**:1267, *1273*
nervous system **9**:1267, *1274*
pancake **9**:1272, *1278–1279*
reproductive system **9**:1267, *1278–1279*
skeletal system **9**:*1272*
spurred **9**:*1279*
toucan **10**:*1404*
trachea (windpipe) **4**:566; **5**:677; **8**:1020; **10**:1374
tracheae
  of invertebrates **1**:107; **4**:*503*; **8**:*1012*, 1015, *1021*–1022
  ants **1**:30
  centipedes **3**:336, 338
  dragonflies **2**:213, 214, 217
  honeybees **4**:*535*
  houseflies **4**:*547*
  lice **5**:*696*
  ventilation **8**:1026

weevils **10**:1353, 1356
tracheids **7**:924; **8**:1144; **9**:1161
tracheoles **8**:1022; **10**:1351, 1356
Tragulina **3**:*356*
transcription and translation **1**:81–*83*; **10**:1320
transmissable spongiform encephalopathy (TSE) **7**:975
transpiration **1**:42
transplants, organ **5**:*617*
transverse processes **9**:*1172*
trapezius muscle **4**:561; **8**:1040–1041; **10**:1370
trees
  evergreen **8**:1140
  rings and dendrochronology **8**:1145
tree shrew **6**:727
Tremarctos ornatus **3**:*410*
Tremella **6**:771
triceps muscles **4**:*560*; **6**:734, *754*; **8**:1040
  of birds **1**:*15*
Trichechidae **5**:*700*
Trichechus
  inunguis **5**:*700*
  manatus **5**:*700*
  senegalensis **5**:*700*
trichobothria **6**:*791*; **8**:1077, 1079; **9**:*1260*
Trichoptera **4**:*538*
Tridacna **3**:343
  gigas **3**:*345*, 347
  maxima **3**:*343*, 344
Tridacnidae **3**:*342*
Trikidae **4**:*460*
trilobite, exoskeleton **9**:1166

trimethylamine oxide (TMO; TMAO) **1**:130; **9**:*1249*, 1250
Triphyophyllum peltatum **10**:1301
Triticum **6**:751, 752
tritocerebrum **2**:215
Triturus **6**:801
Trochilidae **5**:*580*, 581
Trochiliformes **5**:581
Trochilinae **5**:*580*, 581
trochophores **2**:243
trophobiosis **1**:33
trophonemata **9**:1251
tropomyosin **6**:759–760
troponin complex **6**:759
trout **8**:1066, 1067; **9**:**1280–1293**
  brook **9**:*1280*
  brown **9**:*1281, 1284*
  circulatory and respiratory systems **9**:1281, *1289*
  cutthroat **9**:1292
  digestive and excretory systems **9**:1281, *1290–1291*
  external anatomy **9**:1281, *1282–1284*
  gila **9**:1292
  interbreeding **9**:1292
  nervous system **9**:1281, *1287–1288*
  lateral line **6**:*790*; **9**:*1287, 1288*
  rainbow **9**:*1280, 1282, 1283, 1285, 1286, 1290–1293*
  reproductive system **9**:1281, *1292*
  sense of smell **9**:1284, 1287–1288
  skeletal and muscular systems **9**:1281, *1285–1286*
truffle **6**:774, 775, *775*

trumpet hyphae cells **8**:1133
trumpet pitcher **10**:*1300*
Trypanosoma brucei **1**:21
trypanosomiasis, American **5**:622
tsetse fly **4**:546, 548, 549
tuatara, pineal gland (parietal eye; third eye) **2**:273
tube feet **9**:1164, *1227*, *1229*, *1230*, 1231, 1232
tube-nosed seabirds **1**:*10*, 11
  dimethyl sulfide and **1**:16
  evolution **1**:14
  salt glands **1**:18
Tuber **6**:775
tuberculosis (TB) **5**:606
tuberoids **6**:826
tubers, potato **7**:918, *919*, 920, *921–922*, *923*, *925*, 926
Tulasnellales **6**:771
Tulerpeton **9**:1160
tuna **8**:*1060*, 1061
  body temperature **1**:120
  gills **8**:1027, *1071*
  muscles and swimming **6**:764–765; **8**:1066–*1067*
  yellowfin **8**:*1067*
tunicate **3**:*343*; **8**:1025
turbinates, nasal **8**:1031
turtle **5**:*624*; **9**:*1194*, 1195, *1266*, 1267
  big-headed **9**:*1194*, *1266*
  box **9**:1198, 1204, 1268
  breathing **8**:1028
  Chinese softshell **9**:*1201*
  eastern box **9**:1207
  green sea **9**:*1276*

hidden-necked **9**:*1194*, 1195, *1266*, 1273
leatherback, throat **2**:*176*
mud **9**:1198
musk **9**:1198
plastrons **9**:*1198*
pond **9**:*1198*, 1277
sea **9**:*1194*, *1266*
senses **9**:1274
shell **2**:152; **9**:1270
side-necked **9**:*1194*, *1266*, 1273
snapping *see* snapping turtle
softshell **9**:*1194*, *1266*, 1267, 1275
spiny softshell **9**:*1200*
teeth **2**:174
withstanding cold temperatures **9**:1204
tusks **2**:174
  elephant **2**:247–248
twisted-winged parasites **4**:*538*, 542
Tylopoda **4**:*506*; **7**:*964*
tympanic membranes **6**:790
typanosomes **1**:21
typanum **1**:57
typhlosole **2**:242
typhoid **5**:*607*
typhus **5**:*607*

U

Ulmaridae **5**:*636*, 638
ulna
  albatross **1**:*14*
  deer **7**:*970*
  eagle **2**:226
  seal **8**:1116
ulnar sesamoid **3**:298
Ulva lactuca **8**:*1126*
Ulvophyceae **8**:*1126*
ungulates **7**:966; **10**:1361, 1366

even-toed (artiodactyls) **3**:*356*, 357, 377; **4**:*506–507*, *564*; **7**:966; **8**:*1032*; **10**:*1360*, 1361, 1366, *1418*
  piglike **4**:507
  feet **10**:1369
  mammary glands **2**:*275*
odd-toed (perissodactyls) **7**:966; **8**:*1033*; **10**:*1360*, 1361, *1418–1419*, 1421
as relatives of whales **2**:192
unguligrade stance **8**:*1039*, 1041
Uniramia **2**:*208*; **3**:*332*; **4**:*494*; **9**:*1252*
uniramians **2**:208; **3**:*332*; **4**:*494*; **9**:*1252*
United States
  antelope jackrabbits **4**:479
  armadillos **9**:*1181*
  flying squirrels **9**:1213
  manatees **5**:700
  tortoises **9**:*1271*
  turtles **9**:1196
  woodpeckers **10**:*1405*
Unkenreflex **6**:804
urea **1**:62; **2**:186; **8**:1123
Uredinales **6**:771
uric acid **2**:186; **3**:408; **9**:1277
urodaeum **2**:162
Urodela **6**:*800*
urodeum **10**:1416
Urolophus **9**:1247
urophysis gland **2**:267
uropods **5**:687
uropygial gland (preen gland) **1**:13; **10**:1409
urostyle **9**:1160, 1173
Ursavus **3**:415

Ursidae **3**:*410*; **6**:*844*
Ursus
  americanus **3**:*410*
  arctos **3**:*410*
  malayanus **3**:*410*
  maritimus **3**:*410*
Ustilaginales **6**:771
Ustilago **6**:771
uterus **4**:*572*
  bicornuate **3**:372; **6**:*742*; **7**:*945*; **8**:*1046*, 1124; **10**:*1378*, 1403
uteri of kangaroos **5**:*661*; **7**:*996*
utricles **8**:*1133*

V

vaccines, DNA **1**:83
vacuoles
  of amoebas **1**:23
  plant **1**:74
vagina, of kangaroos **5**:*661*; **7**:*996*
Valvatida **9**:*1224*
Vampyromorpha **6**:813
Vampyromorphoida **6**:*812*
vasa deferentia **4**:572–573; **7**:*995*; **10**:1357
vascular bundles **6**:*830*
  cactus **8**:*1055*, 1056
  fern **2**:281–282
  grass **6**:*749–750*
  orchid **6**:829, 831
  Venus flytrap **10**:1307
vasopressin **1**:121
veins **1**:118; **4**:487
velar folds **4**:*455*
Velatida **9**:*1224*
vena cava **8**:1120–1121
Veneroida **3**:*342*
venom glands and venom
  of Cane toads **2**:*275*

venom glands *(continued)*
  of centipedes **3**:333,
    336, 339, *340*
  of honeybees **4**:533
  of jellyfish, coral, and
    sea anemones
    5:640
  of newts and
    salamanders **6**:804
  of octopuses **6**:822
  of scorpions **8**:1081
  of snakes **3**:404
  of spiders **9**:1254,
    1256, 1263
  of stingrays **9**:*1241*
venoms that attack the
  nervous system
  6:783
ventricular fibrillation
  1:117
Venus flytrap
  10:**1300–1313**
  digestive system
    10:1301,
    1308–1311
  external anatomy
    10:1301,
    1302–1305
  internal anatomy
    10:*1306–1307*
  reproductive system
    10:1301,
    1312–1313
Vermilingua **3**:*312*
vertebrae **9**:1172–*1173*
  of anteaters **3**:*320*
  caudal **5**:707; **7**:935;
    9:1173, 1214
  cervical **1**:58; **4**:557;
    6:731; **7**:934;
    9:1173; **10**:1368,
    1410, 1425
  dorsal **5**:654
  of giraffes **3**:363;
    9:*1173*
  of gray whales
    3:381–*382*

of hippopotamuses
  4:512
of humans **4**:557–558
lumbar **4**:558; **5**:654;
  6:732; **7**:934–935;
  8:1038, *1116*;
  10:1410–1411
postsacral **1**:59
sacral **1**:59
of sharks and rays
  9:1172
of snakes **3**:*402*
thoracic **4**:557–558;
  6:731, 732; **7**:934;
  8:1038; **9**:*1172*;
  10:1410, 1425
Vertebrata **1**:*122*; **2**:*148,
  188, 220, 244*;
  5:*700*; **7**:*868, 898*
vertebrates **1**:54; **2**:148,
  244
  heart **1**:*110–112*
  nervous system **6**:781,
    794–799
  responding to stimuli
    6:*782*
Vesicomyidae **3**:*342*
vibrissae *see* whiskers
villi **2**:179, *180*, 182;
  4:*569*; **6**:*741*
viper **3**:401
  heat-detection **3**:406
viral hemorrhagic disease
  (VHD) **4**:488
virions **10**:1314, *1317*
viroids **10**:1317
virus **5**:606;
  10:**1314–1323**
  antigenic drift and shift
    5:621
  classification
    10:1314–1315
  DNA viruses
    10:1314, 1321,
    1322
  immune system and
    5:609, 613–*614*

retroviruses **10**:*1323*
RNA viruses **10**:1314,
  1320, *1321*
structure **10**:1315,
  1316–1318
within cells
  10:*1319–1322*
vision
  bat **3**:297, 303–*304,
    305*
  color **6**:*786*
    dichromatic **9**:1218
    and frugivory **6**:740
    seahorses and **8**:1103
  night-hunting snakes
    3:399
  stereoscopic (binocular)
    1:99; **4**:554; **5**:675;
    6:*736*; **9**:1187;
    10:1396
  underwater, penguins
    7:876
vitamins **1**:72
vitamin B **4**:489
Viveridae **5**:*592, 664*;
  7:*928*
viviparity, in tsetse flies
  4:549
vocal cords **4**:*560*
vole **7**:946
Vombatiformes **5**:648
vomeronasal organ
  (Jacobson's organ)
  3:*406*; **4**:519; **5**:676;
  6:737; **7**:*976*;
  10:*1373, 1395,
  1430*
vulture
  flight **2**:223, 231
  Rüppell's griffon **2**:231

W

wallaby **5**:649, 650
  black-footed rock
    5:*653*
  red-necked **5**:*648*
  rock **5**:649

wallaroo **5**:*648*, 649,
  651
walrus **3**:*410*; **8**:*1110,
  1111*; **10**:*1382*
  blubber **2**:192
  tusks **8**:1125
wapiti *see* red deer
warble fly *see* fly,
  warble
wariness (neophobia)
  7:960
warm-blooded animals
  7:958
warning coloration
  honeybees **4**:530
  newts **6**:804
  snakes **3**:400
  weasels **10**:1330–1331
wasp **1**:*26*, 27; **4**:*538*
  parasitic **1**:35; **4**:*526*
water flea, alternation
  of generations
  7:*989*
water loss, prevention, by
  albatrosses **1**:17
water scorpion **8**:1017
water vascular system
  9:*1229*, 1233
waterwheel plant
  10:1301, 1305
wax glands **4**:533
weasel **3**:*410*; **5**:*592,
  664*; **6**:845; **7**:*928*;
  10:**1324–1343**,
  1325, *1382*
  African striped
    10:1330
  circulatory and
    respiratory systems
    10:*1337–1338*
  digestive and
    excretory systems
    10:1325,
    1339–1340
  external anatomy
    10:1325,
    1326–1331

least **6**:845;
  10:*1324–1327,
  1329, 1332*–1333,
  *1335, 1337,
  1338*–1343
long-tailed **10**:1326,
  1327, 1342
muscular system
  10:1325, *1334*
nervous system
  10:1325,
  1335–1336
northern/snow
  10:1327–1328,
  1340
parasitic worms and
  10:1332
reproductive system
  10:1325,
  1341–1343
short-tailed *see* ermine
size **10**:1340
skeletal system
  10:1325,
  1332–1333
smelly defense **10**:1336
wedge shell **3**:*342*, 343,
  347–348, 351
weevil **4**:528;
  10:1344–1359
bearded **10**:*1351*
black vine **10**:*1359*
chestnut **10**:*1349, 1359*
circulatory and
  respiratory systems
  10:1345,
  1355–1356
cotton-boll **10**:*1346,
  1352, 1355, 1357*
cycad **10**:*1344*
digestive and excretory
  systems **10**:1345
external anatomy
  10:1345,
  1346–1351
fossil **10**:1348
fungus **10**:*1344*

hazelnut **10**:1359
internal anatomy
  10:1345,
  1352–1353
nervous system
  10:1345, 1353,
  1354
nut **10**:1358–1359
pear-shaped **10**:*1344*
pine **10**:*1353*
reproductive system
  10:1345,
  1357–1359
thief **10**:*1344*
whale **4**:*506*
ancient forerunners of
  2:192
baleen **2**:*188*, 189;
  3:*374*–395
relatives **3**:377
beaked **2**:*188*, 190
Blainville's beaked
  3:382
blubber **2**:191, 192
blue **3**:*374*, 377, *379*
calves **3**:395
eyes **3**:*387*
feeding **3**:*392*
stomach **3**:393
bones **2**:194
bowhead **3**:378–379,
  388, 389
Bryde's **3**:*374*
deep-diving **2**:202, 203
DNA collection from
  2:206
fin **3**:*374*, 378
gray *see* gray whale
heat exhanger **2**:202
humpback **3**:*374*, 378
barnacles on **3**:378
hunting by **2**:200
singing **3**:387, 388
skull **3**:*381*
hunting by **2**:200–201
killer (orca) **2**:190;
  3:*377*

communication
  2:201
dorsal fins **2**:190
hunting by **2**:*205*
skull **2**:*195*
teeth **2**:*196*, 204
minke **3**:388
neck **2**:195–196
right **3**:*374*, 375
barnacles on **3**:378
skull **3**:*381*
rorqual, filter feeding
  2:175
senses **2**:200–201
skeletal system, thoracic
  vertebrae **9**:*1172*
skin **2**:191
sperm whales **2**:*188*,
  189
ambergris **2**:205
deep-diving **2**:203
DNA collection from
  2:206
hunting by **2**:200
muscles **2**:197
sexual maturity **2**:207
teeth **2**:204
swimming **2**:198;
  3:376, 379, 384
toothed **2**:*188*, 189;
  3:*374*, 375
brain **2**:199–200
breathing **2**:192
communication
  2:201
digestion by **2**:204
prey **2**:204
reproduction
  2:206–207
senses **2**:200–201
skull **2**:195
wheat **6**:751, 752
whipray, mangrove
  9:1244
whiskers (vibrissae)
  5:703; **6**:848–849,
  855; **10**:*1386–1387*

of lions **5**:674
of pumas **7**:*933*
whisk fern **2**:*276*,
  282
whitefish, northern
  9:1281
wildcat **7**:*928*, 945
wildebeest
  10:1360–1381
black **10**:*1360*
carriers of disease
  10:1374
circulatory and
  respiratory systems
  10:1361, *1374*
common/blue
  10:*1360, 1361,
  1362, 1368, 1370,
  1372, 1374, 1378*
digestive and excretory
  systems **10**:1361,
  1375–1377
external anatomy
  10:*1362–1367*
migration **10**:*1364*
muscular system
  10:*1370*–1371
nervous system
  10:*1372–1373*
reproductive system
  10:*1378–1381*
skeletal system
  10:*1368–1369*
wind scorpion **8**:*1074*,
  1075, *1078*
wings **3**:*299*
bat **3**:294–*295*,
  299–300,
  301–302
dragonfly **2**:211–212
honeybee **4**:529
housefly **4**:544
insect **2**:211
wishbone *see* furcula
Wiskott-Aldrich
  syndrome (WAS)
  5:623

wolf **10**:*1382–1403*
circulatory and
respiratory systems
10:1383, *1397*
communication
10:1388, *1389,*
1401–1402
cubs **10**:*1402, 1403*
digestive and excretory
systems **10**:1383,
1398–1399
Ethiopian **10**:*1382,*
1403
external anatomy
10:1383,
1385–1389
gray **10**:*1382*–1387,
1390–1402
maned **10**:*1382, 1385,*
1391
muscular system
10:1383,
1392–1393
nervous system
10:1383,
1394–1396
packs **10**:1400–1402
red **10**:*1382, 1384,*
1388
regurgitation by
10:1399, *1402,*
1403
reproductive system
10:1383,
1400–1403
skeletal system
10:1383,
1390–1391
teeth **2**:*174*
timber **10**:1385
tracks **10**:*1389*
vocalizations
10:1397
wolverine **3**:419; **6**:845;
10:*1324*, 1325,
1330, 1331, 1335
wombat **5**:*648*

wood **1**:42
soft and hard **8**:1144
wood louse **5**:692
exoskeleton **9**:1166
woodpecker
10:**1404–1417**
acorn **10**:*1408*
avoids brain damage
10:*1413*
black **10**:*1417*
circulatory and
respiratory systems
10:1405, *1414*
crimson-crested
10:*1409, 1415*
digestive and excretory
systems **10**:*1415*
external anatomy
10:1405,
1406–1409
gila **8**:1057; **10**:*1405*
great spotted **10**:*1407,*
1409, 1410, 1414
green **10**:*1416*
hairy **10**:*1406*
Hispaniolan **10**:1416
imperial **10**:1406
making a noise
10:*1414*
muscular system
10:1405, *1412*
nervous system
10:1405, *1413*
pileated **10**:1408–1409,
1410, 1417
reproductive system
10:*1416–1417*
rufous **10**:*1415*
skeletal system
10:*1410–1411*
three-toed **10**:1409
true **10**:*1404*, 1405
Woodsiaceae **2**:*276*
worm-lizards **3**:407
wrack **8**:1127
bladder **8**:*1126, 1128,*
1129, 1131, *1137*

wryneck **10**:*1404*, 1405,
1409, 1415

X
xanthophylls **8**:1127
Xenarthra **3**:*312*;
9:*1176*, 1177, 1191
xenarthrales **9**:1185
xenarthrans (edentates)
3:313, 321
Xiphiidae **8**:*1060*
xylem **1**:40, *65*, 67;
6:829; **7**:924, 925;
8:1056, *1144*, 1145;
10:1306, 1307

Y
yeast **6**:776
reproduction **6**:777
yellow jacket **4**:527
yew
English **8**:1148
plum **8**:1148

Z
Zaglossus **7**:*882*
Zea mays **6**:*749*, 753
zebra **8**:*1032*;
10:**1418–1437**
alertness **10**:1429, *1431*
ancestors **10**:*1421*
avoiding foes **10**:*1427*
circulatory and
respiratory systems
10:1419,
1432–1433
digestive and excretory
systems **10**:1419,
1434–1435
external anatomy
10:1419,
1420–1423
foals **10**:1436, *1437*
Grevy's **10**:*1422*, 1433,
1435
mountain **10**:1419,
1422, 1433

muscular system
10:1419,
1427–1428
nervous system
10:1419,
1429–1431
plains **10**:*1419, 1420,*
1422, 1425–1429,
1432, 1433, *1434*,
1435, *1436*
reproductive system
10:1419,
1436–1437
skeletal system
10:1419,
1424–1426
sounds **10**:1433
stripes **10**:1420–1421,
1422, 1423
Zingiberidae **6**:825
Ziphiidae **2**:*188*
zoospores **8**:1134,
1136–1137
zooxanthellae
in giant clams **3**:342,
343, 352
in sea anemones
8:1092
zorilla (African polecat)
10:*1330*–1331,
1336
zygantra **3**:402
zygapophyses **1**:58;
3:402; **8**:1116
zygodactyl arrangement
10:*1409*
zygomatic arches **7**:952;
9:*1216*
Zygomycota **6**:770, 771
Zygoptera **2**:*208*
zygosphenes **3**:402
zygote **7**:990–991